Civil Engineering and
Engineering Mechanics Series

N. M. Newmark and W. J. Hall, editors

PRENTICE-HALL INTERNATIONAL, INC., *London*
PRENTICE-HALL OF AUSTRALIA, PTY. LTD., *Sydney*
PRENTICE-HALL OF CANADA, LTD., *Toronto*
PRENTICE-HALL OF INDIA PRIVATE LIMITED, *New Delhi*
PRENTICE-HALL OF JAPAN, INC., *Tokyo*

OPTIMUM
STRUCTURAL
DESIGN

OPTIMUM STRUCTURAL DESIGN

LEONARD SPUNT
Associate Professor of Engineering
Department of Mechanics and Materials
San Fernando Valley State College
Northridge, California

Prentice-Hall, Inc., Englewood Cliffs, New Jersey

To my parents

© 1971 by Prentice-Hall, Inc., Englewood Cliffs, New Jersey

All rights reserved. No part of this book may be reproduced in any form or by any means without permission in writing from the publisher.

Current printing (last digit):
10 9 8 7 6 5 4 3 2 1

13-638270-3

Library of Congress Catalog Number: 72-140692

Printed in the United States of America

PREFACE

The reader is no doubt aware of the emphasis placed on analysis in the engineering curriculum. In recent years, however, educators have come to realize that course work in design (synthesis) is becoming a growing necessity. In each specialized application of engineering the design process has taken on its own degree of specialization. Hence, there is a need for studying applications to specific disciplines of engineering as well as the theory of design.

Nevertheless, the predominate theme in the engineering curriculum is still analysis. Its emphasis is justified since a background in analysis constitutes the foundation of any design investigation, and no one would argue with this premise. It should be equally clear, since exposure to the theory of analysis (mathematical modeling, idealization, etc.) is considered insufficient background (in that the student must study analysis in the context of specific application), that exposure to the theory of design is insufficient, and that course work should be provided for its specific application. This text endeavors to bridge the gap between the theory of design and its application to structural systems.

Although background in the theory of design is helpful, it is not required. Chapter 1 presents a complete design methodology for structural applications and does not rely on any design theory prerequisite. The sophomore or junior course work in strength of materials is the minimal background in structural analysis. Course work in stability, although desirable, is not essential. When stability failure modes are considered, the equations will be provided, appropriately discussed, and referenced.

The primary objective of this text is to formulate a unified methodology of optimum design employing parametric analytical procedures in conjunction with numerical algorithm or iterative procedures. The foundations of a unified analytical approach employing "slack variables" are established in Chapter 1 in the perspective of other approaches to the design problem.

Subsequent chapters, by dealing with fundamental classes of load environment, develop an understanding of the relationships between the efficient transmission of these loads and the factors of structural shape and material properties.

It is important to recognize that theory alone in design is basically a contradiction in terms. A clear distinction must be made between the rational processes and the irrational or creative processes that in total we call the design process. In the present treatment this distinction will be in the division of the structural design function into a determination of *form* (shape) and *proportions* (sizes). Once a candidate form has been conceptualized, the optimum determination of proportions is possible by a completely rational procedure. There cannot, however, be a completely rational approach to optimizing form. Here the designer must bring to bear a complement of rationale and creativity in his search for the ideal.

Even when confronted with the more deterministic aspects of design, designers have often been reluctant to employ a parametric optimization method. The typical objections have been that such methods cannot deal with arbitrary constraints on the geometry, systems involving a large number of unknown design variables, multiple (nonconcurrent in time) loading conditions, and the problem of conflicting criteria, e.g., minimum cost and minimum weight. A further objective of this text is to illustrate that a parametric approach can often overcome these objections. Specifically, Chapter 1 introduces a parametric approach for dealing with arbitrary constraints on the geometry with Chapters 2, 4, and 7 illustrating its application to specific examples. In Chapter 3 the question of conflicting criteria is discussed in the context of cost and weight. A "trade-off" criterion is developed here based on an established dollar worth of saving weight. Chapter 5 presents a method for optimizing systems expressed in terms of a large number of design variables while dealing with a mathematical formulation involving only those variables which define the relative orientation of the structure's component elements. The techniques of Chapter 5 rely on closed form parametric optimization results for the component elements of the structure, the development of which is considered in Chapters 2 and 4 for column and beam structural components. Chapter 6 extends the systems approach of Chapter 5 to statically indeterminate structures and multiple load conditions, and demonstrates that redundant loads can be considered independent variables of design.

For the most part, minimum weight will be employed in specific examples as the basis for ranking solutions to the design problem. The emphasis of a minimum-weight-based design is partly a reflection of its importance in aircraft structures and growing importance in civil structures, and partly a recognition that other criteria are seldom explicit mathematical relations which predict merit as a function of the structural geometry. Certainly if the means exist for evaluating the total cost (material, machining, fastening, and other labor costs) in terms of geometry, the approach will apply to a cost-based or trade-off design. For example, the "in place" (material plus labor) costs for steel and concrete have been established for reinforced concrete beam construction on a per unit weight basis. As such, an analytical cost minimization is possible and is developed in Chapter 7 for this case. As another example, the trade-off optimization approach of Chapter 3 is illustrated by an application where total costs (principally fabrication costs) have been determined for given discrete selections of stiffener separation in a stiffened plate. The trade-off optimum is then determined by graphical treatment of this "cost table" in conjunction with an analytical weight function.

The anthor wishes to express his gratitude to many present and former colleagues and students; to Donald Emero and A. L. Kolom for providing a stimulating research environment at North American Aviation, L. A. D., to Professor F. R. Shanley and Professor R. M. Pickett for their valuable discussions on many subjects treated in this book, to Phyllis Osborne and Sergene Zimmerman for their skillful typing of the manuscript, and to my wife for her understanding and constant encouragement.

L. S.

CONTENTS

1
THE METHODOLOGY OF STRUCTURAL DESIGN — 1

1. The Distinction Between Analysis and Synthesis   1
2. The Nature of Structural Design   4
3. The Design Process   6
4. Phase I—Recognition of Environment   7
5. Phase II—Establishment of Criteria   8
6. Phase III—Specification of Form   10
7. Phase IV—Recognition of Constraints   12
8. Phase V—Optimization   13
9. Design Space—Automated Design   20
10. Geometric Constraints   29

2
THE DESIGN OF SLENDER COLUMNS FOR MINIMUM WEIGHT — 32

1. Assessment of the Environment   32
2. The Circular Tube Column   35
3. The H-Section Column   40
4. Weight Index Versus Load Index   45
5. Conditions of End Support   47
6. Geometric Constraints   49

3
DESIGN BASED ON A COST-WEIGHT TRADE-OFF 52
    1 The Dollar Value of a Pound   53
    2 Evaluation of Weight and Cost Merit Functions   54
    3 The Cost-Weight Trade-off Merit Function   55
    4 Trade-off Optimization of Proportions for a Given Form   57
    5 Trade-off Optimization between Alternative Forms   60

4
DESIGN OF SLENDER BEAMS FOR MINIMUM WEIGHT 63
    1 Assessment of the Environment   63
    2 The Circular Tube Beam   64
    3 The H-Section Beam   67
    4 Geometric Constraints   72
    5 Tapered Beams   73

5
DESIGN OF STRUCTURAL SYSTEMS 76
    1 Systems as Combinations of Optimized Elements   76
    2 System Types   77
    3 The Local Environment   79
    4 Analytical Approach to Systems Design   80
    5 Determinate Similar-Element Systems   81
    6 Determinate Dissimilar-Element Systems   87
    7 Geometric Constraints   97
    8 Multiple Load Conditions   97

6
DESIGN OF STATICALLY INDETERMINATE STRUCTURES 102
    1 Redundant External Restraint   102
    2 Redundant Similar-Element Systems   105
    3 Redundant Dissimilar-Element Systems   109
    4 Compatibility   115
    5 Redundant Prestressed Systems   116
    6 General Approach for Trusses under Multiple Load Conditions   126

7
ADDITIONAL PROBLEMS IN CONVENTIONAL STRUCTURAL DESIGN   133
    1   Minimum Weight Wide Columns   133
    2   Frame-Stiffened Cylinder in Bending   139
    3   Minimum Cost Reinforced Concrete Beam   143
    4   Post-Buckling Compression Structure   147

APPENDIX   156
    Selected Material Properties   156
    Material Metrics   157
    Glossary   158

REFERENCES   161

INDEX   165

INTRODUCTION

Many of the engineering structures that exist today evolved to their present level of efficiency by a more or less gradual process in which repeated analysis or iterative procedures played a predominate role. With the advent of the computer, the problem of designing for an optimum state by systematic search of large numbers of candidate solutions has become possible in a reasonable time. Concurrent with the increasing use of the computer in design, the development of sophisticated numerical algorithms for effecting design solutions has evolved in an effort to design systems involving larger and larger numbers of variables.

Although the more complex problems will require some form of computer evaluation, it is the objective of this text to show how certain parametric formulations can complement an algorithmic search method.

An important complement of parametric and search procedures is found in the automated synthesis of large systems employing parametric optimization results for component elements. Chapters 5, 6 and 7 treat the structural systems problem from this vantage point with the first four chapters developing the required component element results. A paramount concern of these component evaluations is the development of relations for the optimum condition of basic structural components (columns, beams, etc.) as a function of their associated loads and lengths. In combination these relations are then employed in subsequent chapters to deal with design questions related to systems of elements. For example, just as component results can establish a ranking between cross sections and material selections for a column as a function of the column load and length, a systems evaluation, employing component results for both a column and a beam, can deal with such questions as: What is the minimum span to depth ratio for a beam bridge structure for which a redundant column support represents improved weight efficiency?

Mathematically the component optimization problem is viewed as a *constrained* minimization with results obtained in parametric form. The function to be minimized represents a goal such as minimum weight, minimum cost or a cost-weight trade-off with constraints representing the suppression of failure by stress or deflection and practical constraints on structural geometry. Utilizing such component results systems evaluations may be cast mathematically as *unconstrained* minimization problems with the number of independent variables reduced to only those which define the relative orientation of component elements and the loads carried by redundant elements.

Most of the applications considered will be minimum weight based designs initially ignoring the effects of practical constraints on geometry. Methods will be introduced at the conclusion of each component application chapter for dealing with such constraints as a modification of the "open variable" solution. Also, a chapter is devoted to the important question of conflicting cost and weight criteria and develops a trade-off approach for such problems.

LIST OF SYMBOLS

A	cross-sectional area	P	concentrated load
C	cost	q	distributed load
c	effective column length ratio	R_N	redundant load variables
D	tube diameter	$R_{U/T}$	ratio of uniform to tapered beam weight
d	concrete beam depth		
E	modulus of elasticity	r	radius of gyration
E_s	fixed system load and length environment	S	system variables
		S_m	material variables
E_{si}	load and length environment for component element	S_o	orientation variables
		S_p	proportion variables
f_e	efficiency function	t	thickness
H	length or depth	V	dollar worth of saving weight
h	web depth in H cross section	W	weight
I	moment of inertia (second moment of area)	δ	deflection
		η_T	tangent modulus ratio
K	buckling coefficient	ρ	weight density
k	redundant load ratio	σ_A	applied normal stress
k_1	ratio of flange width to web depth in H cross section	σ_E	Euler buckling stress
		σ_F	failure stress
k_2	ratio of flange thickness to web thickness in H cross section	σ_f	post buckling failure stress
		σ_L	local buckling stress
L	length or span	σ_y	yield stress
M	merit function or moment	σ_{PL}	proportional limit stress
M_{cw}	cost-weight trade-off merit function	ψ	slack variable

OPTIMUM STRUCTURAL DESIGN

1
THE METHODOLOGY OF STRUCTURAL DESIGN

Design is one of the primary functions of engineering. In designing, the engineer creates a method, device, process, or, more broadly, a system, the objective being to satisfy a performance requirement while minimizing those factors which reduce the efficiency of the system.

The creative act in engineering design is not an act of creation in the sense of the Old Testament. It does not create something out of nothing; it uncovers, selects, reshuffles, and combines. In short, it *synthesizes*. Synthesis is the rational approach in design by which the engineer manipulates an "existing body" of facts, ideas, skills, equations, and other information to achieve the design objective. This existing body of information represents the *analysis* capability for the fixed and variable quantities in the design with respect to their influence on performance and efficiency, and accordingly provides the foundation for a rational approach to design. Without this capability, design is at best an empirical or evolutionary process. Actually most of the early engineering achievements were accomplished in the absence of the applicable analysis foundation. The erection of the pyramids, the Roman highways and arches, and the uniformly stressed bow are just a few examples of the countless engineering systems created long before their underlying principles became quantitative tools. In the age of modern engineering, empirical and evolutionary design have become supplemental to the rational approach of synthesis.

Section 1

THE DISTINCTION BETWEEN ANALYSIS AND SYNTHESIS

The distinction between analysis and synthesis is fundamentally one of viewpoint or objective while employing a common body of knowledge. Analysis represents the capability to predict the performance of a physical system by manipulation of a mathematical model whose behavior in the abstract reproduces that of the idealized real system. Synthesis simply represents the shift of viewpoint that, given the ability to predict the performance of a system, one can manipulate to predict those systems which manifest a given performance. Since there is no unique system which exhibits a required performance, synthesis must further be based on an established measure of efficiency so that the multiple "system solutions" can be ranked accordingly. The following figure represents, in block diagram, a rational approach to design illustrating the distinction between analysis and synthesis. Note the reliance on analysis capability and a design objective defined in terms of a performance requirement and a measure of efficiency.

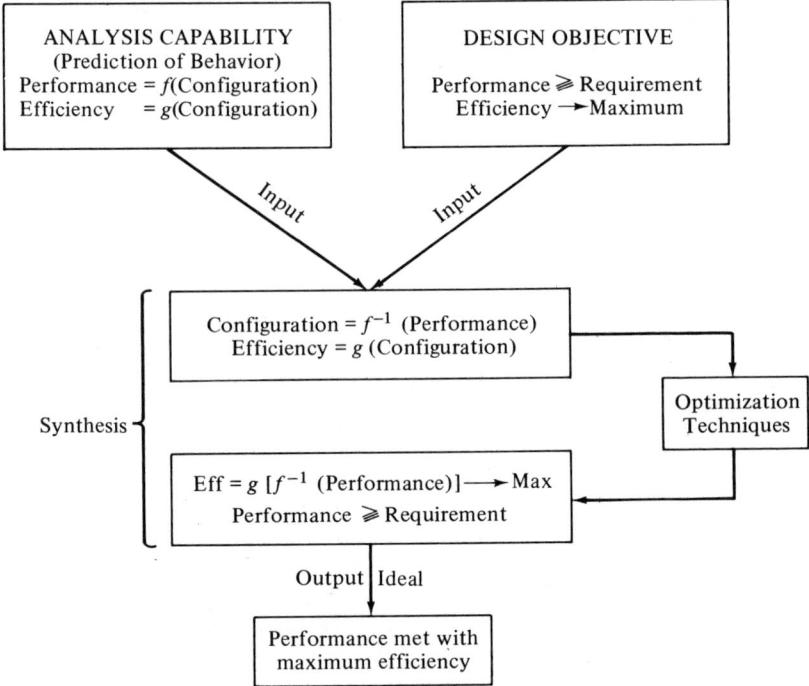

Fig. 1-1 (a). Rational Design by Synthesis Based on an Analysis Capability and a Design Objective

Between the inputs and the output ideal, there are three blocks which represent the application of a synthesis procedure in general symbolic terms. The first block implies a process of "inverting" the analysis capability of performance as a function of configuration so that the configuration becomes the dependent quantity. This relation, combined with efficiency as a function of configuration, is manipulated within the block labeled optimization techniques until the efficiency is expressed in terms of the performance requirement as shown symbolically in the third block. The efficiency is then maximized, subject to the performance requirement as a constraint.

As a specific example, consider the simple design problem of finding the thickness of a pressurized open-ended (no longitudinal stress) cylinder with fixed radius. The performance requirement is that no yielding should occur and efficiency is established as a minimum in material volume. Fig. 1-1(b) shows the corresponding block diagram.

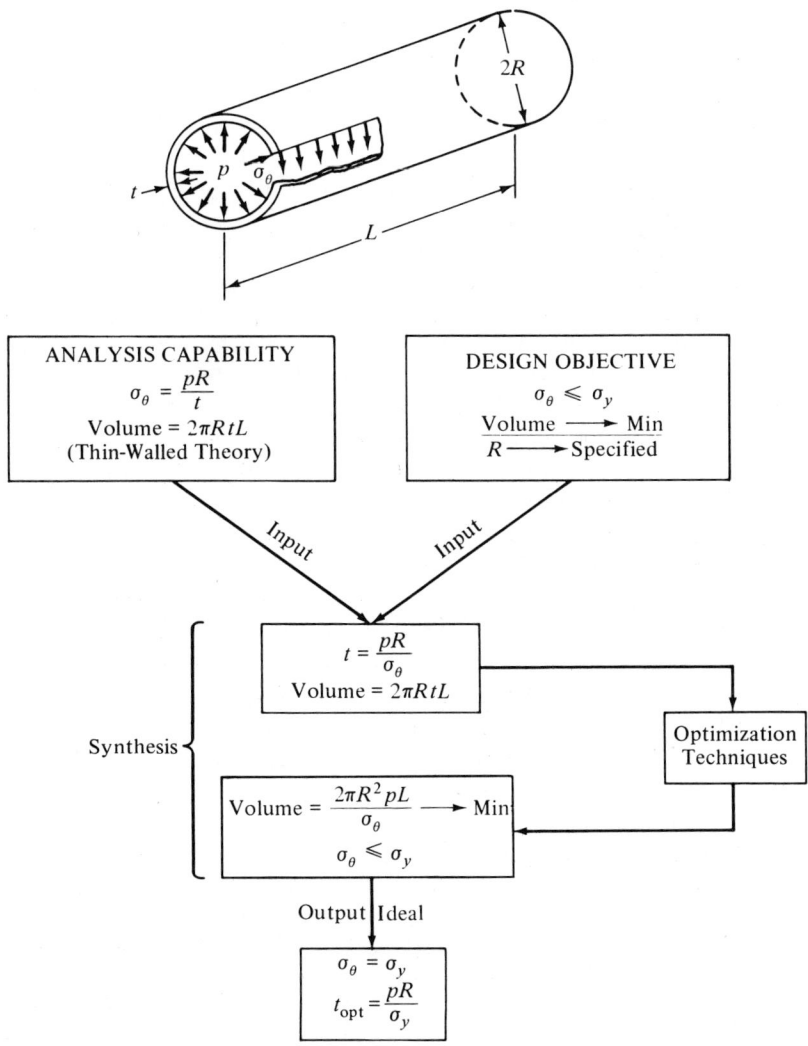

Fig. 1-1 (b). A Simple Example

It should be emphasized that these block diagrams are presented for the purpose of illustrating the distinction between analysis and synthesis. They are not meant to be an exhaustive exposition of a synthesis methodology. What is needed is a systematic formulation of the numerical and analytical procedures within the optimization techniques block, a goal we shall presently pursue in the perspective of a broader design methodology.

Section 2
THE NATURE OF STRUCTURAL DESIGN

Structural design is the process of determining the *configuration* (*form* and *proportions*) of a structure, subject to a load-carrying performance requirement and based on an established measure of efficiency. The form of a structure describes the *shapes* and relative arrangements of its component elements. The proportions are the quantities which define *size* for the component elements and for the structure as a whole.

A structure can be described broadly as a material body whose primary function is to contain internally and/or support externally other bodies while sustaining loads which result from the dynamic transport or static bearing of these bodies. The important point to be gleaned from this description is that the primary function of a structure is not to sustain loads. The load-carrying requirement is generally the quantitative representation of containment and support requirements. As a consequence, the best form of a structure is not necessarily that which most efficiently sustains the load environment.

Even in those cases where there are no restrictions on form, they must be imposed in an iterative fashion so as to reduce the design problem to a determination of proportions. Without any restrictions limiting the parameters that define form, there is nothing substantial for which equations can be written. In particular, the variables which define size cannot be expressed literally, nor can the equations which express the inherent failure modes be written, unless the structural form has at least been specified. Consequently, the determination of an efficient form is basically a trial-and-error procedure with the result that no assurance can be had that the final form is *absolutely* optimum (in essence an *open-ended* problem).

Although there can be no rational a priori approach which predicts the absolute optimum form for a given load environment, if we limit our focus to a specific class of alternatives, the evolution of an optimum form becomes a relative determination and, as such, an a priori approach is possible. An important illustration of this is found when the class of forms is limited to axial force transmission elements excluding buckling effects. A priori approaches to form optimization for minimum weight under these limitations have been developed by Michell[1] based on earlier theorems derived by Maxwell.[2] It is important to recognize that here Michell has restricted the solution to a class of forms for which most of the parameters of form have already been set. Specifically, for this *truss* configuration, the proportion variables can be expressed literally in terms of the cross-sectional areas, A_i, of each element, and the manner in which these elements are mutually fastened, combined with the ignoring of buckling effects, results in an a priori knowledge of a common failure mode (excessive stress) for each element.

Such evaluations are still of considerable interest since they provide the absolute lower bound weight for a truss network under a given load environment. As conventional structures, however, these solutions are seldom practical in that they yield an excessive number of elements which result in undergauge and excessive fastening weight. These diminished returns would seem inherent in any a priori approach to form optimization. We recognize, however, that such "bounding" evaluations can establish important frames of reference in a systematic methodology.

In the absence of lower bound references, optimum configuration can only be established between the immediate alternatives. Even prior to analytical evaluation, the notion of *qualitative assessment* can be applied to those forms for which an analysis capability exists. We shall see in subsequent chapters how such assessments can narrow down the candidates considered for follow-up analytical optimization to those which can reasonably expect to result in high efficiency (e.g., the observation that a truss system should be more efficient than a beam, since axial force—which fully stresses the cross section—is inherently more efficient than bending moment transmission).*

The theory of structural design, therefore, concentrates on the determination of the optimum proportions of a variety of form types or classes for given load-carrying functions and on the establishment of trends resulting from these investigations. Although many forward steps have been taken toward this desired goal, all too often the procedure in engineering practice is iterative, with the factors of adequate strength (performance) and efficiency established by analysis,† and with no assurance that the endpoint is optimum. Although we must generally accept this lack of assurance in the case of form determination, there is no need to accept it in the determination of proportions.

---

*It is shown in Chap. 5 that this is only true for a specific range of load environments. At extremely high intensity load environments the beam structure is more efficient.

†The results of this strength and efficiency analysis permits the next guess to be more "educated" than the previous one. We shall refer to this approach as *iterative design*.

Section 3
THE DESIGN PROCESS

After Thomas Edison had investigated one hundred material combinations for the filament of the electric light bulb, his laboratory assistant expressed frustration, claiming that he could see no sign of progress in nearly a hundred failures. Edison replied, "We are making exceptional progress. We now know one hundred possibilities that won't work."

What Edison was really saying is simply that the probability of success increases with the number of failures. The stockpiled information on why a large number of design solutions are inadequate provides the very basis for formulating a better solution.

In the context of structures, it is the manner in which various structural forms fail under the given loads that provide the insights from which an improvement evolves. For example, suppose that we wish to determine the least volume of a slender element to carry a given bending moment. If we focus our attention on a solid shape with, say, circular or rectangular cross section, we find that the least volume solutions to these form types fail such that practically all of the material is understressed. Only the material fibers with maximum distance from the bending axis are stressed to their full potential. As such, a tubular cross section, which eliminates much of this unused material, is indicated as an improvement. The wide beam of fixed width and depth (the aircraft wing) provides a more intriguing illustration of this process. Here, as a result of the fixed width and depth restriction, even the tubular shape has shortcomings since, at optimum conditions, the compression surface will generally buckle at a stress considerably below the material's full potential. Again the inadequacies of this form type provide clues to a more efficient solution. The compression surface could be stiffened by employing attached or integral stiffening elements or a full depth egg crate web or rib supporting core. Using such stiffening, the fixed depth form can be optimized so that buckling occurs at a higher stress level. The fact that such a form is more efficient than a simple tube, however, is not an a priori determination in that it has to be arbitrated as a candidate *before* this can be shown. Although this is not so much arbitrated as it is implied by a discovered inadequacy of a more primitive form, the important point is that it evolves as an efficient form based on prior alternatives and is not deduced as such from the considerations of load-carrying requirements and efficiency measure alone.

We leave now the question of form optimization (we will have occasion to return to it in Chaps. 5 and 6 in the context of specific systems). Our immediate objective is to formalize the design process as a systematic "step" procedure so that for some "arbitrated" structural form the optimum value of each of the proportion variables can be determined as a function of a given load environment.

The systematic approach which effects the conceptualization of candidate forms and the ranking of their relative efficiencies for some prescribed performance requirement is termed the *phases of design*. It should be noted that many authors have dealt with the design process in engineering as a step procedure. Such phrases as "The Morphology of Design," "The Anatomy of Design," and their inherent systematic methodologies have found wide application in engineering practice.[3]

In the present treatment, the phases of design will be developed for the specific application of structural design problems.

Section 4
PHASE I—RECOGNITION OF ENVIRONMENT

The initial phase of design is a recognition phase in the sense that it implies a process of recognizing a structural load environment and the performance factors which impose limitations on structural shape. This process may seem self-evident and not necessarily an integral part of a systematic approach in design. Subsequent phases of design, however, are highly dependent on a clear and detailed recognition of the environment. In general, this recognition can involve any or all of the following factors:

1. A structural design problem exists for which a literature search produces no adequate solution.

2. The problem manifests itself in a set of predicted loads and their respective spatial and time coordinates.

3. There may be existent material geometry in the vicinity of the load environment which can "support" structural forms.

4. There may be access, containment or airfoil requirements, which preclude the selection of certain structural forms and materials.

The quantitative results of this phase of design is an array of parameters which describe the systems environment and, as such, we define them as *system environment* and denote them as an array of quantities by E_S. The qualitative aspects of the environment we term *environmental factors*, e.g. arifoil requirements (see Fig. 1-2(d)).

It should be emphasized that, although this and subsequent phases of design will be presented as separate and distinct steps, there will generally be overlapping. For example, in some cases the recognition of environment cannot be completed until a structural form is specified. A specific example of this would be in the case of wind loads.

Fig. 1-2 shows several examples of structural load environments.

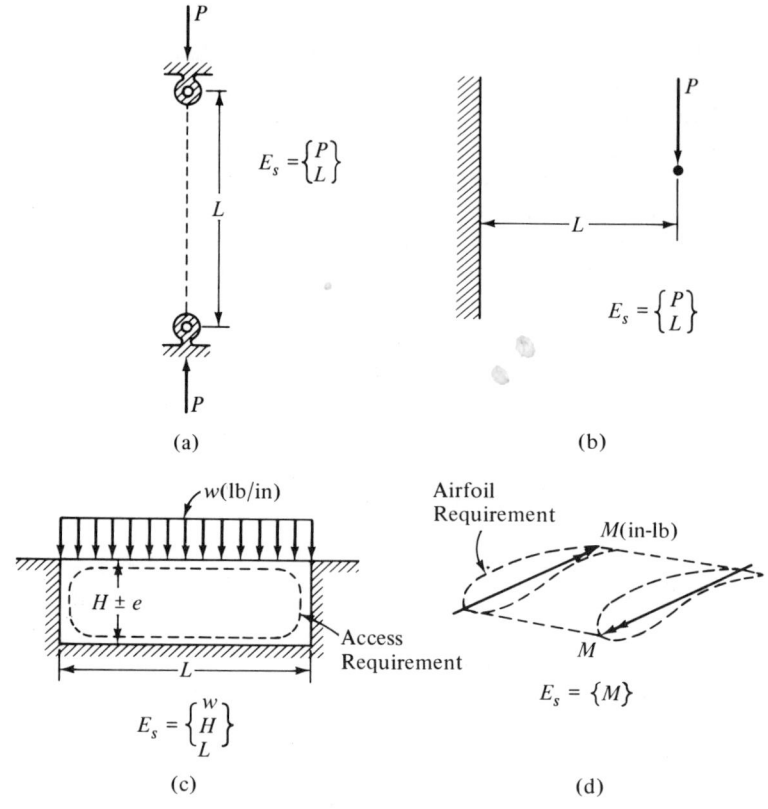

Fig. 1-2. Examples of Recognized System Environment

Section 5

PHASE II—ESTABLISHMENT OF CRITERIA

In design, the term optimum denotes the attainment of an absolute maximum (or minimum) in one or more measurable quantities (or qualities). Since the establishment of the factors we seek to optimize involves value judgments, an optimum state must be taken as that which exists in the eye of the beholder and not as an intrinsic characteristic of the performance requirement.

The term "best" is often used synonymously with optimum. Although there may be subtle differences in that optimum implies an absolute state whereas best may imply a comparison yielding one better than specific alternatives, practically speaking there is little difference. The state of the art imposes limitations for which optimum, as an applied concept, must be interpreted as best relative to a class of alternatives. It is in this context that the term optimum will be employed in this text. Accordingly we made the following definition:

> An optimum structural configuration (form and proportions) is the best, on some established basis, of those candidates of a class of alternatives which are acceptable under the constraints.

What constitutes acceptability will be considered in Phase IV (Recognition of Constraints) and, although it will limit the candidates for the singular title of best, it does not influence the basis for which they can be compared for this title. The basis for comparing (or ranking) we term *criteria*. The criteria defines a measure of value and accordingly gives us the ability to make a "decision" between any two candidate design configurations. Although this is a much-sought-after ideal, it is difficult to attain practice. Often the expression of criteria is self-contradictory. For example, the desire to have both minimum weight and minimum cost can be a contradiction. The process of compromising between two (or more) contradictory requirements in the criteria is referred to as an establishment of a *trade-off*. For example, a cost-weight trade-off might be expressed as follows: "The weight should be minimum, but no more than $100 should be expended for each pound of weight saved."

Hopefully the principal part of the criteria or an appropriate trade-off can be expressed analytically in terms of a configuration-dependent function, where the configuration which renders this function an extreme is optimum. This function, if it exists, is termed the *merit function** and denoted by M. In structural applications the merit function is normally weight, cost, or a cost-weight trade-off.

The weight of a structure (with the exception of fastening weight) can always be written as a function of the geometric dimensions once a structural form has been specified. The cost, however, may be difficult to express analytically in these terms. In some cases, the cost is principally material cost and, consequently, cost minimization reduces to weight minimization. If the cost is principally production expense (machining, fastening, etc.), the cost merit function is usually empirically established, resulting in a table of values for particular candidates. In Chap. 3, a method will be developed for the minimization of a composite merit function based on a cost-weight trade-off criterion.

The establishment of criteria is usually the result of an interpretation of the requirements and goals of an engineering system which must contain load-carrying structure. Although there is a certain amount of feedback, the structures engineer seldom has any say in the establishment of criteria. His

---

*It can also be referred to as criterion function or object function.

role is simply to be aware of the criteria and to apply them systematically in the design of the structural components of the system. The Apollo program, for example, by virtue of its mission requirements and the limitations set by existing boosters, required extremely light-weight structures and the principal criterion was accordingly established as weight. Most commercial aircraft, on the other hand, not being motivated by such lofty ideals, are designed structurally on the basis of a cost-weight trade-off established by an assessment of weight's effects on mission performance and projected profit expectations.

In addition to the principal criterion which manifests itself in the merit function, there are other *qualitative* criteria which, although not expressed analytically, will nevertheless influence subsequent phases of design and in particular the specification of form. These are the qualities we seek to maximize. Examples would be aesthetics, safety, and production feasibility.

Section 6

PHASE III—SPECIFICATION OF FORM

This phase of the design process, perhaps the most critical, is the most difficult from the standpoint of applying an analytical procedure. The search for perfection in form is an open-ended characteristic of design for which no analytical procedure can hope to corral and reduce to equation. To synthesize form, the designer utilizes his ability to predict the structural behavior of various form types under the given load state and his prior experience with similar design problems. We recognize, however, that such a process can provide only clues to the best solution, not its complete identity.

Above all, the selection of alternative candidate forms should be based on a logical appraisal of the nature of the environment and criteria. Certainly, if the criterion is principally cost, a form favoring a minimum number of easily machined elements would be indicated, whereas if the criterion is principally weight, the reverse might be indicated. It must also be stressed that a background in the analysis of a variety of structural elements is an important prerequisite, not only to this particular phase but to the design process in general.*

After due consideration to the above-mentioned factors, there may be several alternative forms which, from a *qualitative assessment*, are indistinguishable. In this case, each should be optimized and the results compared quantitatively.

---

*The simple observation that the more efficient way to carry a load in an enclosure, by cable supporting it from the ceiling as opposed to cantilevering it from a wall, would be far less straightforward in lieu of the quantitative ability to analyze the states of stress under axial and flexural loads.

Once a form has been specified, idealized, and a mathematical model constructed, a set of variables is defined which describes the proportions of the various elements, their relative orientation, and their material characteristics. Accordingly we define three types of *system variables* denoted as arrays of quantities by S_p, S_o, and S_m, and referred to respectively as *proportion variables*, *orientation variables*, and *material variables*. The proportion variables define the various dimensions of the specified form such as thicknesses, widths, and lengths of flat plate elements, radii of curved plates, distance between stiffeners in a stiffened plate, angles between plates in a corrugation, depth of core in a sandwich plate, etc. The orientation variables define the spatial configuration of a system of elements in such terms as number and location of nodes in a frame gridwork, angles between truss members, distance between column supports in a bridge structure, etc. The material variables simply define the mechanical properties of available materials such as modulus of elasticity, yield stress, density, etc. Note that each structural element (referred to as an *element* or *subsystem*) will have its own distinct pair of variable arrays S_p and S_m but that there will only be a single variable array S_o for the total structural form (referred to as a *system*). To illustrate this, consider Fig. 1-3, which shows a combination of tension cable and compression tube as a candidate form for the environment of Fig. 1-2(b).

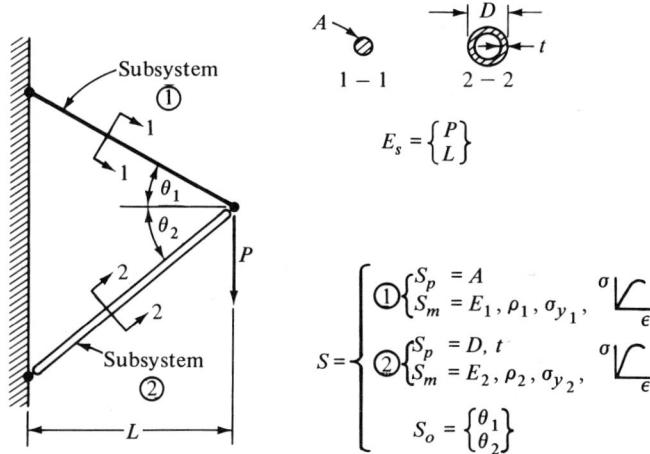

Fig. 1-3. Example of Specified Form

The system variables can be further classified into *primary* and *secondary* variables. We define primary variables as those whose optimum values depend on a numerical expression of the environment and secondary variables as those whose optimum values, although dependent on the nature of the environment, are independent of a numerical expression of the environment; e.g., in Chap. 7 the least-weight corrugation wide column yields an optimum corrugation angle of 30°, independent of load or transmission length magnitudes.

It is not necessary, at this stage, to read intrinsic meaning into the classifications of parameters and variables that are being made within the phases of design. At this point, they can be considered arbitrary. As we proceed with specific applications, their utility will become evident.

Section 7

PHASE IV—RECOGNITION OF CONSTRAINTS

If we consider the system variables as axes in an n-dimensional hyperspace, then each point in that *design space* represents a candidate design. The question of which of any two points is the better design can be answered by substituting the respective coordinates into a configuration-dependent merit function and comparing. This comparison, however, is only realistic if both points represent *acceptable* designs. A design point is considered acceptable if, under the given environment, no inherent failure modes occur, e.g., yielding, buckling, fracture; also, the respective system variables do not violate established limits, e.g., minimum required sheet thickness, maximum permissible beam depth, minimum required tube diameter.

The insistence that acceptable design points correspond to structural configurations for which all inherent failure modes are suppressed is termed satisfaction of *failure constraints*. These constraining relations are a set of *inequalities* for each subsystem which express the condition that the applied stresses must not exceed the stresses for which failure occurs and occasionally express minimum stiffness requirements. A given failure constraint is generally a function of the element's proportion and material variables, the orientation variables, and the systems environment. We can denote these constraints generally by $FC(S, E_s) \geq 0$, or specifically in terms of stress failure suppression as $\sigma_A \leq \sigma_F$.*

The insistence that acceptable design points contain no coordinates which violate arbitrarily established limits is termed satisfaction of *geometric constraints*. We denote these constraints generally by $LB \leq GC(S) \leq UB$, where LB and UB are respectively the lower and upper bounds. If, for a given system variable, $LB = 0$ and $UB = \infty$, the variable is unconstrained (in a geometric sense) and accordingly is termed an *open variable*. In most cases

---

*In what follows, we take σ_A to mean an applied stress, e.g. P/A for an axially loaded member, and σ_F to mean a failure stress, e.g., the yield stress for the material.

the function $GC(S)$ reduces to a particular system variable of the form S_p or S_o. It can, however, involve a function of several system variables (see Fig. 1-4(b)). Fig. 1-4 illustrates examples or arbitrarily established geometric constraints.

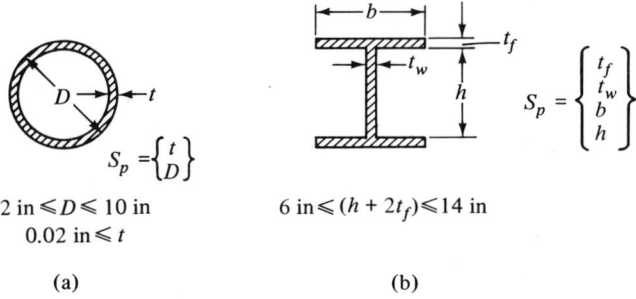

2 in $\leqslant D \leqslant$ 10 in   6 in $\leqslant (h + 2t_f) \leqslant$ 14 in
0.02 in $\leqslant t$

(a)   (b)

Fig. 1-4. Examples of Geometric Constraints

Although the geometric constraints are arbitrarily established, the failure constraints, on the other hand, are implied by the nature of the environment's action on a specified form. They are not established, arbitrarily or otherwise. The principal role of this phase, therefore, involves a detailed *recognition* of these inherent failure modes. The completeness of this recognition must hinge on the extent of the designer's background in structural analysis.

Section 8
PHASE V—OPTIMIZATION

This phase involves the techniques of solving a mathematical formulation of Phases I through IV, shown below as Eqs. (1-1) through (1-5).
  For a recognized environment

$$E_s \tag{1-1}$$

and a specified structural form

$$S \begin{cases} ① \begin{cases} S_p \\ S_m \end{cases} \\ ② \begin{cases} S_p \\ S_m \end{cases} \\ \vdots \\ S_o \end{cases} \tag{1-2}$$

S will be the optimum set $\{S\}_{opt}$ if
$$M(S, E_s) \longrightarrow \min\ (\max) \qquad (1\text{-}3)$$
subject to the failure constraints
$$\begin{aligned} FC_1(S, E_s) &\geq 0 \\ FC_2(S, E_s) &\geq 0 \\ &\vdots \end{aligned} \qquad (1\text{-}4)$$

and the geometric constraints
$$\begin{aligned} LB_1 &\leq GC_1(S) \leq UB_1 \\ LB_2 &\leq GC_2(S) \leq UB_2 \\ &\vdots \end{aligned} \qquad (1\text{-}5)$$

The term optimization denotes the various techniques of maximizing merit (Eq. (1-3)) subject to the *inequality* constraints of Eqs. (1-4) and (1-5). These techniques can be classified broadly as either numerical or analytical.

Numerical Methods

The use of systematic numerical algorithms is termed *automated design* and involves the branch of mathematics known as *mathematical programming*.[4] This approach is beyond the scope of this text, but for perspective will be discussed briefly in Sec. 9 of this chapter. Suffice it to say here that automated design is a numerically "directed-search" procedure through the acceptable design candidates. It is typically accomplished on a computer since it involves exhaustive calculation. Also, it effects a so called "point solution" by evaluating the optimum design for a specific numerical expression of the environment.

Analytical Methods

An analytical technique relies on the algebraic combination of Eqs. (1-3) and (1-4), where the inequalities of Eq. (1-4) are, by some appropriate subterfuge, converted to equations. The advantage of such an approach lies in the *parametric* evaluation of M_{opt} as a function of the environment and material properties. This parametric capability permits the ready evaluation of the most efficient material, facilitates the comparison between forms for the general environment, and also reduces the optimization of multiple element systems to a determination of orientation variables and redundant loads

(Chap. 5). The disadvantage of this approach is in the difficulty of directly satisfying the geometric constraints (Eq. (1-5)). In many cases these constraints, if not too numerous, can be accounted for by an indirect analytical procedure.*

Simultaneous Mode Design

A widely used analytical technique in structural applications is *simultaneous mode design*† (SMD). The technique incorporates a principle—over and above the merit function—for the determination of an optimum design (normally with respect to minimum weight). The principle can be stated as follows:

> A given form will be optimum if all failure modes which can possibly intersect occur simultaneously under the action of the load environment.

The pioneering application of this principle is attributed to F. R. Shanley[5,6]. Shanley's analytical optimization of columns shows clearly the parametric advantage of a merit function evaluated in terms of material properties and load environment (which he terms structural index) and accordingly provides a cornerstone for the theory of optimum structural design.

Obviously the technique of SMD is an "appropriate subterfuge" for conversion of the failure constraint inequalities into equations. When these equations are combined with the merit function, subsequent mathematical optimization will automatically satisfy (identically) the failure constraints, and the design thus determined will correspondingly be acceptable.

Lagrangian Multipliers

If in applying the SMD technique the failure constraints cannot be explicitly combined with the merit function due to unwieldy algebra, optimization is still possible, employing the Lagrangian multiplier technique for determining a stationary value subject to satisfaction of conditional *equalities*. The Lagrangian technique is presented below without proof (for a proof see any advanced text in applied mathematics).

---

*By indirect it is meant that the flexibility or "slack" inherent in the failure constraints and "excess" proportion variables can be employed to satisfy geometric constraints. (This will be discussed in Sec. 1-10.)

†It is also termed "One Hoss Shay" design after Oliver Wendell Holmes' legendary carriage that was built so well that it lasted "one hundred years to the day" then fell apart all at once.

The determination of set

$$x_1, x_2, \cdots$$

which renders

$$f(x_1, x_2, \cdots) \longrightarrow \min\,(\max) \qquad (1\text{-}6)$$

subject to satisfaction of

$$g_1(x_1, x_2, \cdots) = 0$$
$$g_2(x_1, x_2, \cdots) = 0 \qquad (1\text{-}7)$$
$$\vdots$$

can be effected by minimizing (maximizing) the function

$$H = f + \lambda_1 g_1 + \lambda_2 g_2 + \cdots \qquad (1\text{-}8)$$

with respect to $x_1, x_2, \cdots \lambda_1, \lambda_2, \cdots$ (the λ are termed the Lagrangian multipliers).

Slack Variables

In this text we will employ an analytical technique which avoids the occasionally restrictive* technique of SMD. The technique employs *slack variables* (denoted by the symbol ψ) as a subterfuge for converting failure constraint inequalities to equations while still *retaining* the inequality nature of the constraints. This is accomplished by subjecting the slack variables themselves to inequality constraints of the form $\psi_i \leq 1$. The failure constraints, as equalities containing the slack variables, can then be combined with the merit function whose optimization can be effected parametrically with respect to the slack variables. The limitations imposed by the constraints, $\psi_i \leq 1$, can then be considered in light of their direct parametric influence on the optimization of the merit function. This technique will be presented below as a mathematical generality for the most common type of failure constraints, those which express the insistence that the various applied stresses (σ_{A_i}) must not exceed the inherent failure stresses (σ_{F_i}).

S will be the optimum set $\{S\}_{\text{opt}}$ if

$$M(S, E_s) \longrightarrow \min\,(\max) \qquad (1\text{-}9)$$

---

*It is shown in Chap. 4 that a family of optimum cross sections exist for the H-cross section under bending. Although one member of this family is in fact the SMD, the generality of the family solution is lost using the SMD approach. In Chap. 7 it is shown that the minimum cost reinforced concrete beam is generally not a SMD. Also, the SMD technique is generally limited to an "open-variable solution," i.e., it cannot satisfy geometric constraints.

subject to the constraints

$$\sigma_{A_1} = \psi_1 \sigma_{F_1}$$
$$\sigma_{A_2} = \psi_2 \sigma_{F_2}$$
$$\cdot$$
$$\cdot$$
$$\cdot$$
(1-10)

where

$$\psi_1 \leq 1$$
$$\psi_2 \leq 1$$
$$\cdot$$
$$\cdot$$
$$\cdot$$
(1-11)

The reader should compare Eqs. (1-10) and (1-11) with Eq. (1-4). Note that they are equivalent in the sense that either Eqs. (1-10) and (1-11) or Eq (1-4) impose inequality failure constraints. The difference lies in the manner in which these constraints have been expressed. In Eq. (1-4), the suppression of inherent failure modes has been written directly as constraining inequalities. In Eqs. (1-10) and (1-11), the constraining inequalities have been expressed indirectly as a set of conditional equations involving slack variables, all of which are subject to the same constraining inequality, $\psi_i \leq 1$. This results in the ability to combine the conditional equations with the merit function, in which case subsequent optimization will automatically satisfy the conditional equations. The constraints on the slack variables, however, must still be contended with. On the surface this may appear to be a process of "diminished returns," resulting in no apparent advantage. The advantage exists in the simplicity and uniformity of the ψ-inequalities and, more important, in the ability to optimize the merit function as a parametric relationship *containing* the slack variables.

To illustrate the technique of analytical optimization, consider the following simple problem in structural design.

EXAMPLE 1-1. Derive an expression for the minimum weight of a tensile member under the environment of Fig. 1-5(a). Generalize with respect to weight's dependence on material selection (see Appendix) and environment. We shall specify as a candidate form a solid cylindrical member as shown in Fig. 1-5(b). The proportion variable is simply the cross-sectional area (A). The pertinent material variables are density (ρ) and yield stress (σ_y). In this simple example the recognition of environment is illustrated in Fig. 1-5(a). The criteria is established in the problem statement and yields the following merit function

$$M(S, E_s) = \rho A L \longrightarrow \min \quad (1\text{-}12)$$

From an assessment of environment and criteria the specification of form as illustrated in Fig. 1-5(b) is a logical one. Under axial tensile stress there is no buckling problem which, if present, would necessitate a more efficient distribution of cross-sectional area (a hollow tube, for example, in order to provide an efficient bending rigidity).*

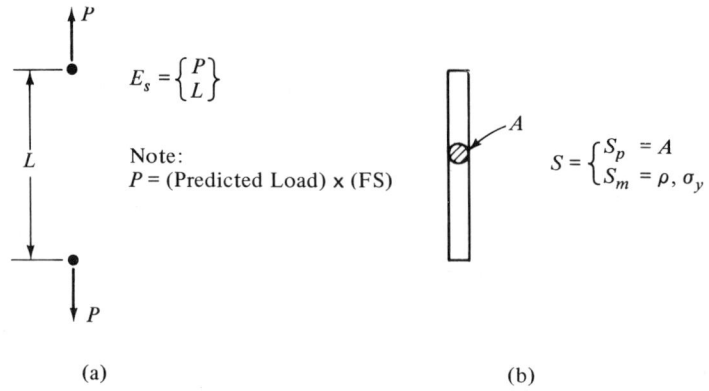

Fig. 1-5. Environment and Form for a Tensile Member

We next consider the recognition of inherent failure modes. For the case of tensile stress this depends on whether failure is defined as yielding or actual material separation (fracture). In most design applications yielding is taken as failure since exceeding this condition results in excessive (irreversible) deformations. Accordingly, the single failure constraint becomes

$$\sigma_A = \psi \sigma_y, \quad \psi \leq 1$$

or

$$\frac{P}{A} = \psi \sigma_y \qquad (1\text{-}13)$$

subject to

$$\psi \leq 1$$

The general technique proceeds by combining Eq. (1-13) with Eq. (1-12), in the direction of reducing the number of unknown proportion variables in the merit function. Since in this example there is only one such variable (A), the result is that the weight can be written as a function of environment and material variables with parametric dependence on ψ as follows.
From Eq. (1-13)

---

*A more efficient distribution of cross section would also be indicated if a minimum transverse stiffness requirement were desired.

$$A = \frac{P}{\psi \sigma_y} \qquad (1\text{-}14)$$

Substituting this into Eq. (1-12) yields

$$M = \frac{\rho P L}{\sigma_y \psi} \qquad (1\text{-}15)$$

subject to

$$\psi \leq 1$$

Obviously, considering the possible values of ψ which correspond to acceptable designs ($\psi \leq 1$), the optimum value is $\psi_{\text{opt}} = 1$. Hence,

$$M_{\text{opt}} = \frac{\rho}{\sigma_y} PL \qquad (1\text{-}16)$$

Eq. (1-16) represents the optimum weight as a function of material and environment. The optimization is completed by evaluating the system variables as similar functions. In this example, substituting ψ_{opt} into Eq. (1-14) yields

$$A_{\text{opt}} = \frac{P}{\sigma_y} \qquad (1\text{-}17)$$

The generalizations requested are evident from Eq. (1-16). They are:
1. The most efficient material will be that for which the quantity σ_y/ρ is a maximum (see Appendix).
2. The optimum weight is directly proportional to both P and L. If the loading requirement is doubled, for example, the weight will double.

If the optimum values of the slack variables are all respectively unity (as in the example above), the design thus determined is, by definition, a simultaneous mode design. It is important to note that, in this manner, the result is "post factum" in the sense that it was deduced as opposed to assumed as a principle in the design. Although the application of the SMD principle would have yielded identical results in the above example, there are design situations where the SMD technique adopted "a priori" will be restrictive.

The reader should recognize that Example 1-1 is presented in order to illustrate the technique of analytical optimization employing slack variables in the simplest possible manner. A complicated problem at this stage might obscure the otherwise straightforward steps applied in this technique. The versatility of this approach can only become apparent in its application to more complicated structural elements involving two or more proportion variables and two or more inherent failure modes. We will consider numerous design problems of this nature in subsequent chapters.

Before doing so, however, in order to consider analytical optimization in its proper perspective, we will first consider, in such detail as commensurate with the scope of this text, the numerical approach of automated design.

Section 9

DESIGN SPACE—AUTOMATED DESIGN

It was pointed out in Sec. 7 that a hyperspace can be imagined for which each point in that space represents a possible design. Generally the space is constructed (in a mathematical sense) for the proportion and orientation variables (S_p, S_o) with specified numerical values for the material variables (S_m) and the system environment (E_s). We can also imagine the failure and geometric constraints as hypersurfaces which divide design space into acceptable and unacceptable regions. Another type of hypersurface, vital to the concept of design space, is the family of constant-value contours of the merit function. These are surfaces for which each point therein contained yields a constant value for the merit function. As this constant is changed a family of such surfaces is generated. The optimization approach termed automated design can best be described through a graphical visualization of design space. Restricted as we are to a two-dimensional medium of communication, a graphical interpretation of a hyperspace is beyond reach; we can, however, graphically display a design space involving two system variables. This planar space is shown in Fig. 1-6 for a hypothetical design problem where the merit function (M) is to be minimized.

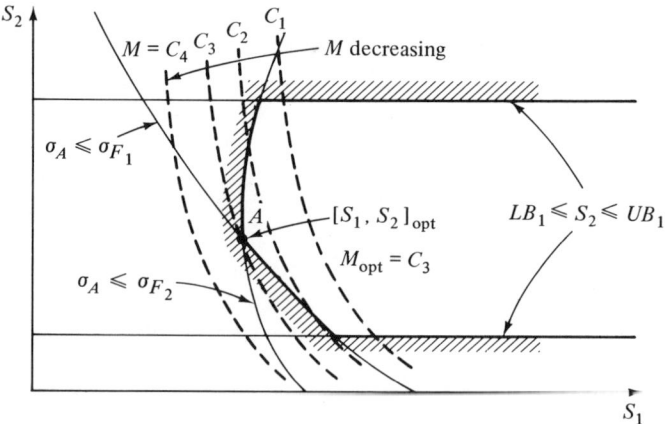

Fig. 1-6. Graphical Visualization of Design Space

The acceptable design region is bounded by a composite curve defined by the various failure and geometric constraints.* By constructing a family of constant-value contours of the merit function, shown in the figure as dashed curves, the optimum design can be seen to be point A, corresponding to the merit function $M = C_3$. This is true since for contours such that $M < C_3$ ($M = C_4$, for example), no design point therein contained falls within the acceptable design region. It should be noted that only a single point on the optimum contour is in the acceptable region and, consequently, a unique optimum design with the coordinates of point A is found corresponding to an absolute minimum for M at C_3. Also, this point occurs at the intersection of the failure constraints which means that the design point is a simultaneous mode design. This is often the case for optimum designs which fall within the geometric constraints and is the principal motivation for the application of the SMD technique in open variable design.

The fact that optimum designs which do not fall within the geometric constraints cannot be evaluated using the SMD technique can be illustrated by increasing the lower bound constraint (LB_1) to a value in excess of $S_{2\text{opt}}$. The design space which results is shown in Fig. 1-7. Note that point B now becomes the optimum design due to point A violating the geometric constraint on variable S_2. Although point B corresponds to $\sigma_A = \sigma_{F_2}$, it can be seen that σ_A does not equal σ_{F_1} (the failure mode corresponding to σ_{F_1} is said to be conservatively suppressed) and therefore the optimum design, by virtue of invoking a geometric constraint, is not a simultaneous mode design.

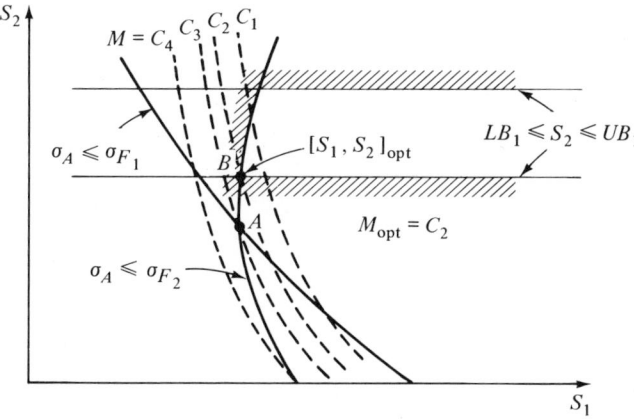

Fig. 1-7. Effect of a Geometric Constraint

*The unacceptable region is shown crosshatched.

To describe fully the techniques of automated design would require mathematical techniques beyond the scope of this text. For a sophisticated mathematical treatment see Ref. 7. Examples of its application to building design and the waffle grid compression plate can be found in Refs. 8 and 9. We will content ourselves here with a qualitative description of two of the techniques of automated design: (1) the method of *steepest descent* and (2) the method of *random steps*. The main concern of this text is with analytical optimization employing slack variables. To grasp the potential of any method, however, is to be able to put it in perspective with other approaches to the design problem. It is toward this end that we explore the nature of automated design and not in the hope of developing the applicable mathematical tools.

The Method of Steepest Descent The evaluation is made through a computer program which contains the failure and geometric constraint functions and the merit function. The material variables as well as the system environment must be numerical inputs to the program. The technique is essentially a directed sequential movement through design space beginning with an input design point well within the acceptable region of design (it is a relatively easy matter to propose a highly conservative design). Program execution begins by evaluating numerically the direction in hyperspace for which movement from the initial design point will most rapidly improve the merit function. This preferred direction* is obviously along the normal to the merit contour which contains the initial design point (see Fig. 1-8). Sequential steps are taken until either a constraint surface is encountered or the merit function (evaluated at each step) fails to improve. If the merit function fails to improve obviously we have found (at the very least) a relative optimum (referred to as a local optimum). Such a local optimum, however, may not be the absolute optimum sought after. There is no test, as yet, which will guarantee an optimum to be absolute. One way to reach a level of confidence that a proposed optimum is absolute is to repeat the evaluation several times, beginning with initial design points widely separated. If each time the design converges to the same local optimum, there is good reason to believe that this is the absolute optimum sought after. Let us return now to the other possibility that, although the merit function continually improves, the direction of steepest descent eventually intersects a constraint surface. Obviously, to proceed further in this direction would force the design point into the unacceptable region of design. When this occurs the computer evaluates a modified direction parallel to the constraint surface (so as not to violate it) and in a direction which again causes the most rapid improvement in the merit function. This is accomplished mathematically by removing the component of the original steepest descent which is normal to the constraint surface. This new direction

---
*Referred to as the direction of steepest descent.

(called a constrained steepest descent) is followed until again either the merit function fails to improve or another constraint surface is encountered. The process of direction modification continues until a subsequent design point fails to improve the merit function within some predetermined degree of accuracy.

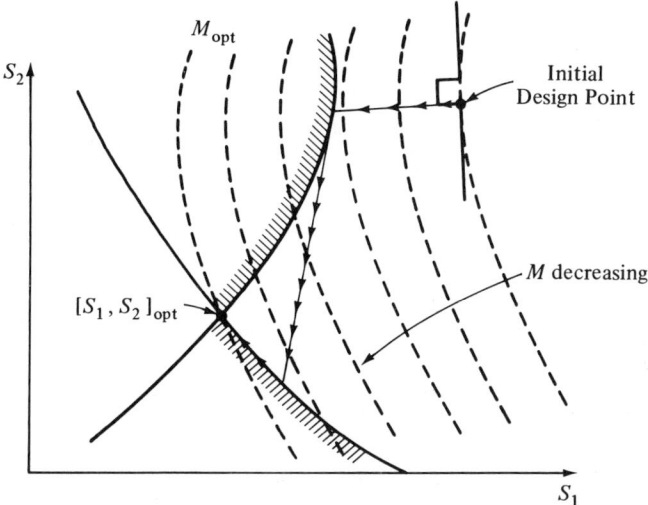

Fig. 1-8. Visualization of Steepest Descent

*The Method of Random Steps** This method, although similar to the method of steepest descent in that it also involves repeated direction modification, requires far less mathematical programming since each new direction is determined by random chance; the only requirement being that the direction thus determined at least causes the merit function to improve. The degree of improvement, however, is not considered as it was in the case of steepest descent where maximum improvement was guaranteed by insisting that the new sequential movement be in the direction of the contour normal. The resulting lack of efficiency is compensated for by the relative ease in which the computer program can be written. In cases where the method is applied, the increased execution time is happily tolerated since a steepest descent approach would have involved insurmountable mathematics. The method is illustrated graphically in Fig. 1-9. Again, execution begins with an initial design point well within the acceptable region of design. A random number generator is employed in order to establish the direction of design improvement, discarding those directions (if they happen to come up) which obviously

*Also referred to as the Monte Carlo method.

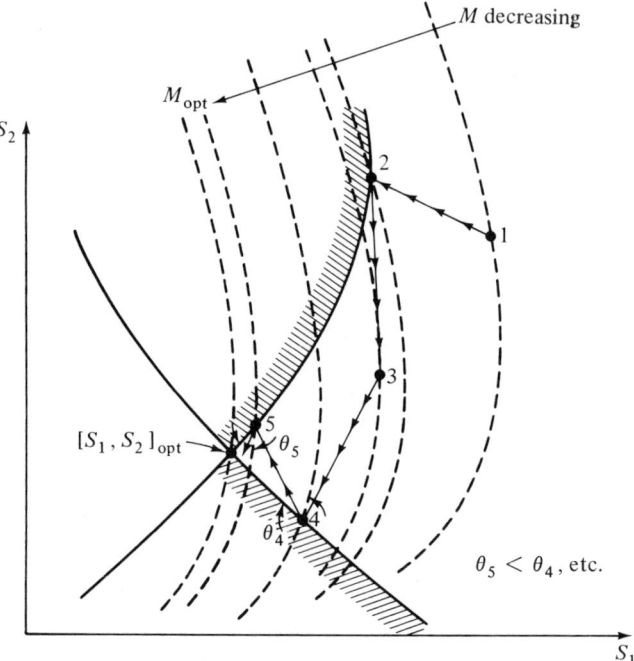

Fig. 1-9. Visualization of Random Steps

will not improve the merit function. This is accomplished by advancing one step in the random direction and if the merit function fails to improve the direction is replaced with another randomly determined direction. The random determination of an acceptable direction is obviously a 50–50 proposition and in a few trials a modified direction is found. Sequential steps are taken in this direction until either the merit function fails to improve or a constraint surface is encountered. Unlike the method of steepest descent, if the merit function fails to improve, a local optimum has not necessarily been located and a new direction must be evaluated (note point (3) in the figure). The new direction is again randomly determined under the restriction that the merit function improves. If a constraint is encountered the modified direction is subject to the additional restriction that the design should not move across the constraint into the unacceptable region. The number of trials required to yield an acceptable random direction is accordingly increased. When a point is reached where a large number of trials fails to turn up an acceptable direction, there is a high probability (commensurate with the number of trials) that the current design point is very close to a local optimum. This can be seen in the figure by comparing point (4) with point (5), and noting the tendency for the angle between the contour tangent and the constraint tangent to become successively smaller as the design approaches an optimum.

From the above descriptions it should be evident that the techniques of automated design evaluate the optimum for a given numerical expression of the environment and the material variables. The inability to evaluate an optimum design as a function of the environment may not, on the surface, seem like a serious drawback (isn't the environment fixed anyway?). In the optimization of systems, however, a given structural element (subsystem) may be subjected to a subsystem environment which will depend not only on the fixed system environment (E_s) but also on the orientation variables (S_o).*
An advantage of subsystems optimized as functions of their environments, therefore, is realized in subsequent system's optimization where the respective environments cannot be numerically fixed since they contain the orientation variables which are themselves subject to optimization.

The advantage of automated design is found in ability to deal with "unwieldy" geometric constraints and problems too mathematically complex to be solved by analytical means.

It must be emphasized that, besides being the theoretical basis for automated design, the concept of design space can be a powerful graphical tool for structural problems involving two proportion variables. Though it can only be constructed for given numerical values of material and environment, the relative orientation and respective shapes of the constraint and merit curves can, nevertheless, yield valuable generalizations about the optimum design and also insights which might reduce the labor in developing a general analytical optimization solution.

EXAMPLE 1-2. Repeat the design evaluation of Example 1-1 using a graphical design space. Take for the environment $P = 10,000$ lb and $L = 100$ in, and use AISI 1025 steel. In addition to the inherent failure mode of yielding, assume that it is required to limit the maximum transverse deflection, due to the member's own weight, to 0.1 percent of its length. This transverse stiffness requirement is to be evaluated for simple supports at the end points, in a horizontal configuration and under zero axial load.
In light of the transverse stiffness requirement we will specify a hollow tube as the structural form since, from the standpoint of bending rigidity, it represents a more efficient distribution of cross-sectional area then that of a solid rod. As shown in Fig. 1-10(a), we have for the axes of design space D and t. The pertinent material variables are now ρ, σ_y, and E and for the given material their values are respectively 0.283 lb/in^3, 36,000 psi and 30,000,000 psi.

*Refer to the system depicted in Fig. 1-3.

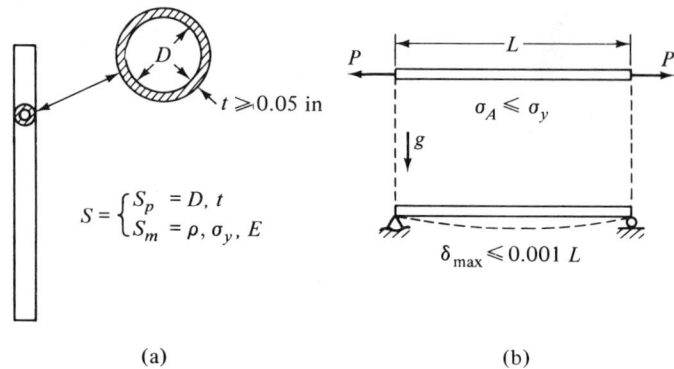

Fig. 1-10. Specified Form and Recognized Constraints

The merit function, again weight, can be expressed for the tube configuration (using a thin-walled approximation) as

$$M = \rho \pi D t L \longrightarrow \min$$

which for the given values of ρ and L, becomes

$$88.8 \, Dt \longrightarrow \min \tag{1-18}$$

The failure constraints, as shown in Fig. 1-10(b), are next expressed in terms of system variables and environment. For the yield constraint we have

$$\frac{P}{\pi Dt} \leq \sigma_y$$

which for the given values of P and σ_y, becomes

$$D \geq \frac{0.0885}{t} \tag{1-19}$$

For the deflection constraint we have

$$\delta_{max} = \frac{5qL^4}{384EI} \leq 0.001L$$

q is the uniformly distributed transverse load, and can be expressed as $\rho \pi Dt$ for a tube under its own weight. For thin-walled tubes, the moment of inertia (I) can be expressed as $\pi D^3 t/8$. Substituting for q and I, and introducing numerical values for E and L yields

$$D \geq 0.994 \tag{1-20}$$

Since in design space we can deal directly with geometric constraints, we shall establish a limitation on the thickness as follows

$$t \geq 0.05 \tag{1-21}$$

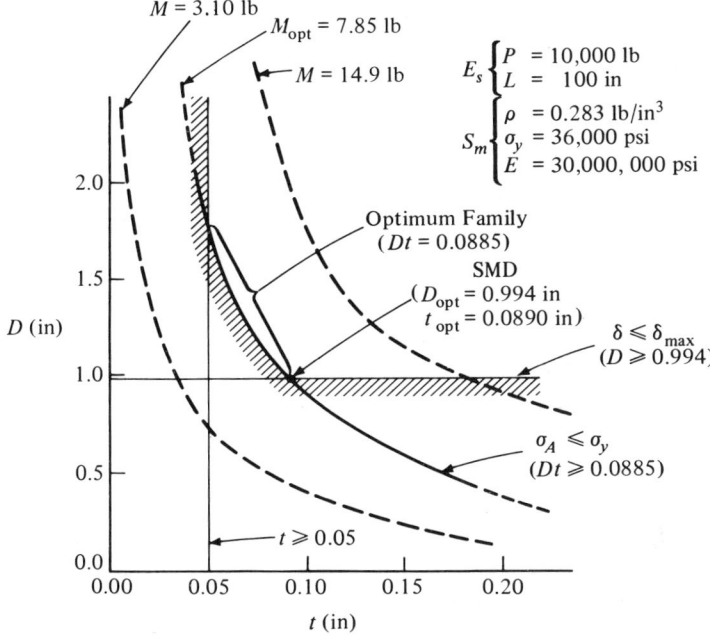

Fig. 1-11. Design Space for Example 1-2

The resulting design space, shown in Fig. 1-11, is constructed by first establishing the composite constraint contour. Equation (1-19), for example, is satisfied by plotting the hyperbola equality and then noting that for a given t, D must be "greater than or equal to" the value predicted by the equality. Therefore all points below the equality curve are unacceptable designs and accordingly this region is shown cross-hatched. Similar arguments are applied to Eqs. (1-20) and (1-21) and the composite constraint curve is thus determined (shown in bold line). From Eq. (1-18), the constant value merit contours are a family of hyperbolas, one of which, therefore, must be coincident with the yield constraint curve. As a result, this contour can be seen to be the optimum, corresponding to $M_{opt} = 7.85$ lb. Note that an infinite number of optimum pairs of D and t can be found on the optimum contour within the constraints (referred to as a family of optimum designs). This family* can be expressed as

$$88.8\, D_{opt} t_{opt} = 7.85$$

provided

*One of which is the simultaneous mode design.

and
$$t \geq 0.05$$
$$D \geq 0.994$$

The above example illustrates the possibly restrictive nature of the SMD technique. Although the design space has demonstrated that the SMD is optimum, it is obvious that had the analytical SMD technique been employed the generality of the family solution would have been lost.

EXAMPLE 1-3. Construct a two variable design space and determine $\{S\}_{opt}$ for the following merit function and constraints. Show also that an identical optimum can be obtained employing a slack variable analytical solution

$$M = S_1 + S_2 \longrightarrow \min \qquad (1\text{-}22)$$

subject to

$$S_1 - 2 \geq 0 \qquad (1\text{-}23)$$
$$S_1 + 2S_2 - 6 \geq 0 \qquad (1\text{-}24)$$

The merit and constraints all plot as straight lines as shown in Fig. 1-12. Again the unacceptable region is shown cross-hatched with merit contours shown dashed.

From the design space it is seen that the optimum point is $S_{1_{opt}} = S_{2_{opt}} = 2$. Note that this point corresponds to $M_{opt} = 4$ and the intersection of constraints. For an analytical solution we introduce slack variables into Eqs. (1-23) and (1-24) yielding

$$S_1 - 2 = \psi_1 \qquad (1\text{-}25)$$
$$S_1 + 2S_2 - 6 = \psi_2 \qquad (1\text{-}26)$$

subject to

$$\psi_1 \geq 0, \quad \psi_2 \geq 0$$

Solving the above for S_1 and S_2 in terms of ψ_1 and ψ_2 gives

$$S_1 = \psi_1 + 2 \qquad (1\text{-}27)$$
$$S_2 = \frac{1}{2}(\psi_2 - \psi_1) + 2 \qquad (1\text{-}28)$$

Now eliminating S_1 and S_2 from Eq. (1-22) by substituting Eqs. (1-27) and (1-28) results in

$$M = \frac{1}{2}(\psi_1 + \psi_2) + 4 \qquad (1\text{-}29)$$

Since ψ_1 and ψ_2 must be non-negative we conclude that $\psi_{1_{opt}} = \psi_{2_{opt}} = 0$ yielding a minimum for M of 4. Inserting the optimum values of ψ into Eqs. (1-27) and (1-28) results in $S_{1_{opt}} = S_{2_{opt}} = 2$.

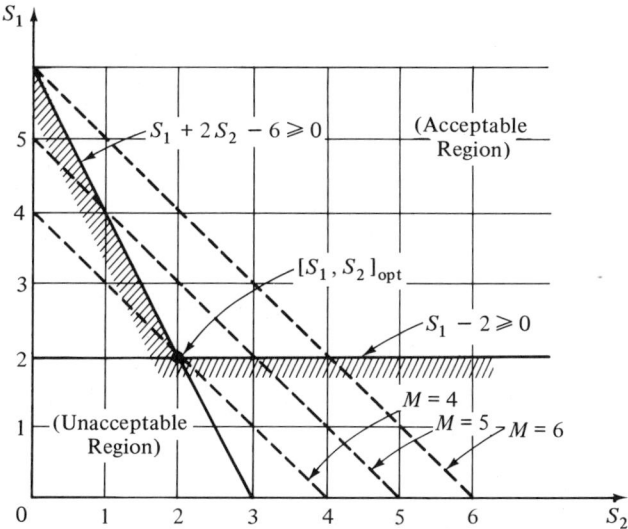

Fig. 1-12. Design Space for Example 1-3

Section 10
GEOMETRIC CONSTRAINTS

Depending on the number and type of failure constraints and proportion variables, it is often possible to express the merit function in terms of slack variables alone, in which case the open variable optimization problem reduces to a search for $\{\psi_i\}_{opt}$ with the proportion variables being dependent evaluations. Since here all proportion variables and merit are dependent on ψ_i, it is possible to effect geometric constraints by an appropriate selection of $\{\psi_i\}_{co}$ (constrained optimum) such that particular geometric constraints are satisfied while minimizing the associated merit penalty. This approach to geometric constraints will be illustrated in Chap. 2 for a tubular column subject to a minimum thickness constraint and in Chap. 4 for a tubular beam subject to a fixed diameter constraint.

In those cases where the merit function cannot be expressed in terms of slack variables alone, the merit function must be optimized with respect to some of the proportion variables in addition to slack variables. These *excess proportion variables*, if expressed as ratios or angles, will generally be secondary system variables in an open variable solution. In a geometrically constrained evaluation, however, it is these excess variables which, by parametric adjustment, can generally be employed to effect a given geometric constraint with minimum merit penalty. This approach to geometric constraints will be illustrated in Chap. 7 for a flat corrugation panel as a wide column.

It is important to recognize that a constrained optimum condition can only be realized if *all* proportion variables are evaluated subject to the geometric constraints. Even if only one proportion variable is explicitly constrained, to optimize under an open variable soultion and then arbitrarily change only the variable that is constrained will prevent a "redistribution of material" which could change all of the proportion variables. To illustrate this, refer to Fig. 1-7. Note that the geometric constraint on S_2 causes S_1 to be changed from its open variable optimum in order to achieve the constrained optimum condition. In subsequent applications of analytical optimization, the algebraic equivalent of the construction of Fig. 1-7 will be effected employing the "degrees of freedom" to move through design space afforded by slack variables and excess proportion variables. (In a multiple-element system, the orientation variables afford additional degrees of freedom in a geometric constraint evaluation. This will be discussed in Chap. 5.)

PROBLEMS

1-1. Construct the equivalent of the block diagram illustrated in Fig. 1-1(a) using the symbolic terminology of Eq. (1-9) through (1-11) and words appropriate to the application of structural design.

1-2. List the advantages and disadvantages of the following design techniques.
1. Empirical and evolutionary design
2. Iterative design
3. SMD
4. Slack variables
5. Automated design

1-3. Sketch candidate forms for the following environments (as many as you like). Define $\{S_p\}$ for each subsystem you employ, and $\{S_o\}$ for the total system form.

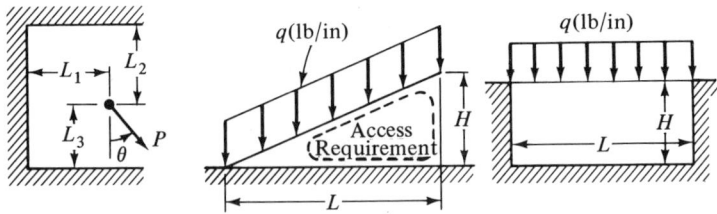

1-4. Using an analytical optimization approach work Example 1-2 for the general environment, P, L, and material variables, ρ, E, σ_y. Compare this result to the "design space" solution for the specific environment and material of Example 1-2.

1-5. Show that E/ρ is the measure of the weight efficiency of a material as a tensile member subject to an axial deflection limitation only. Assume elastic behavior.

2

DESIGN OF SLENDER COLUMNS FOR MINIMUM WEIGHT

A slender column is any structural configuration which sustains the environment of Fig. 2-1.

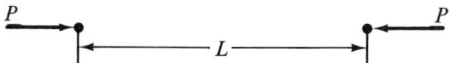

Fig. 2-1. Slender Column Environment

Even before a form is specified, we can conclude that this manner of force transmission, regardless of the form which supports it, will impose the inherent failure mode of general column instability (Euler buckling). We begin, therefore, by considering the functional relationship between environment and configuration geometry expressed by this failure constraint in the hopes of providing some insight into what might be an efficient form. Any inherent failure mode implied by the environment alone should be considered at the outset so that the arbitrariness of form specification can be minimized.

Section 1

ASSESSMENT OF THE ENVIRONMENT

The general column instability constraint can be expressed in terms of the well-known Euler critical buckling load as follows.*

$$P \leq P_E$$

where

---

*cL in Eq. (2-1) represents an effective column length, where c depends on the end supports. For pinned ends $c = 1$. See Sec. 5 for other end conditions.

$$P_E = \frac{\pi^2 \eta_T EI}{(cL)^2} \qquad (2\text{-}1)$$

or

$$\pi^2 \eta_T EI \geq P(cL)^2 \qquad (2\text{-}2)$$

Since the product PL^2 is fixed by the environment, Eq. (2-2) expresses a minimum requirement for the flexural rigidity (EI) of any specified form regardless of its shape.*

In order to satisfy Eq. (2-2) in the direction of minimum weight, it will be necessary to specify a form for which the minimum value of EI† will be large compared to the cross-sectional area. The question of the best form of a cross section for maximum EI per unit area is fundamental to design problems involving the buckling or bending of slender members, and accordingly we shall consider it in some depth.

For the case of maximum EI per unit area about a specified bending axis (as in the case of a given bending moment), the ideal is achieved by locating the bulk of the area a significant distance from the bending axis and providing a web shear tie as shown in Fig. 2-2.

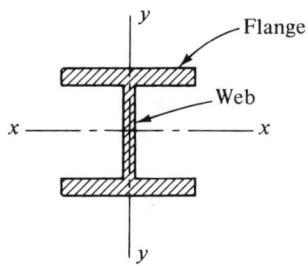

Fig. 2-2. The H-Section as a Candidate Form

Although EI per unit area is large about the x–x axis, it is not so for bending about the y–y axis since, in this case, the web makes no contribution to I and the flanges are not as effectively located with respect to the bending axis. As a minimum-weight column cross section, therefore, this form is not going to be a strong contender, since the "path of least resistance" would obviously be buckling about the y–y axis. From the standpoint of production expense, however, this form may be more efficient than alternative forms

---

*Being mainly concerned here with a qualitative assessment, we postpone the consideration of the variations in the tangent modulus ratio (η_T), which for inelastic behavior becomes a stress-dependent function.

†A buckling column takes the path of least resistance and hence buckles about the centroidal bending axis for which EI is least. It is this minimum value which must be employed in Eq. (2-2).

since it can be mill rolled with practically no machining or fabrication expense. The fact that a given form may not be the absolutely minimum weight configuration does not imply that we should ignore its relative weight efficiency. Subsequent cost-weight trade-off studies can only be accomplished in light of the respective candidates' weight efficiencies together with their cost and producibility factors.

Returning now to the question of possible improvements in the cross-sectional distribution of Fig. 2-2 so that the minimum value of EI per unit area is large, it is evident that a logical selection towards this end would be either a circular or square tube, as shown in Fig. 2-3.

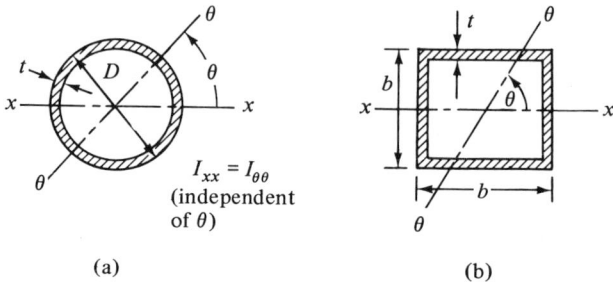

Fig. 2-3. Circular and Square Tubes as Candidate Forms

These particular cross sections have the property that the moment of inertia about any centroidal bending axis is invariant. We can refer to these cross sections as *balanced* in the Euler mode. From the standpoint of suppressing Euler buckling, this property is desirable in that if a given axis produced a larger value of I than another, the greater value would represent extraneous material distribution since buckling will occur about the axis for which I is least. A decision at this point as to whether the circular tube or the square tube represents the most efficient form would be stretching the application of a qualitative assessment of environment beyond its potential. It should be recognized, however, that with the specification of these two candidate forms, the inherent failure modes associated with their respective cross sections could be assessed. For example, both candidates, being tubular, are susceptible to *local buckling* modes. The circular tube is more efficient in this mode than the square tube (curved sheet requires more buckling load than flat sheet) and intuitively the circular tube might therefore seem a more optimum form. Although this is the case, the reader is cautioned against making such conclusions before the fact, especially in the case of close contenders. In many design problems intuition can be misleading, and should be considered a supplement to analytical evaluation and not a substitute.

Section 2
THE CIRCULAR TUBE COLUMN

Consider the cross section of Fig. 2-3(a). Employing the technique of analytical optimization as outlined in Eqs. (1-9) through (1-11), we first establish the governing merit function as

$$W = \rho AL \longrightarrow \min \tag{2-3}$$

and, in so doing, restrict the solution to a uniform value of A (D and t) along the column length.*

The failure mode of Euler buckling will be expressed in terms of stress by dividing both sides of Eq. (2-1) by A and yields

$$\sigma_A \leq \sigma_E$$

where

$$\sigma_E = \frac{\pi^2 \eta_T E}{\left(\frac{cL}{r}\right)^2} \tag{2-4}$$

(r is the radius of gyration, defined as $r = \sqrt{I/A}$).

Introducing a slack variable into Eq. (2-4) results in

$$\sigma_A = \psi_E \frac{\pi^2 \eta_T E}{\left(\frac{cL}{r}\right)^2} \tag{2-5}$$

$$\psi_E \leq 1$$

Under axial compression a circular tube has an inherent failure mode which results in a "local" diamond pattern of buckled waves as shown in Fig. 2-4.

The governing failure constraint is given by

$$\sigma_A \leq \sigma_L$$

where

$$\sigma_L = K_c \eta_T^{1/2} E\left(\frac{t}{D}\right) \tag{2-6}$$

Introducing a slack variable, the local buckling constraint becomes

$$\sigma_A = \psi_L K_c \eta_T^{1/2} E\left(\frac{t}{D}\right) \tag{2-7}$$

$$\psi_L \leq 1$$

---

*Feigen[10] has shown that a tapered circular tube is at most 11 percent lighter than the uniform case. Excessive stress at the ends (which taper to zero) would reduce this figure somewhat.

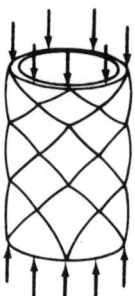

Fig. 2-4. Local Buckling of a Circular Tube

In Eq. (2-7), the form of the inelastic correction, $\eta_T^{1/2}$, is as recommended by Shanely in Ref. 6, p. 18. The theoretical value of the buckling coefficient, K_c, is 1.212 (Ref. 11). Due to finite material imperfections, however, a much lower value is found in actual tests. These tests indicate that a conservative value would be $K_c = 0.40$ (Ref. 6).

The third and last failure mode is excessive stress. In engineering design, the maximum permissible normal stress is usually taken as yield (σ_y). Accordingly the failure constraint can be written as

$$\sigma_A \leq \sigma_y \tag{2-8}$$

We shall see that due to the following conversion of the merit function to an applied stress form, a slack variable will not be necessary for Eq. (2-8).

The general approach requires combining the failure constraints, Eqs. (2-5), (2-7), and (2-8), with the merit function, Eq. (2-3), in the direction of reducing the number of unknown proportion variables. Since the constraints are all written in terms of the applied stress, it will be convenient, if possible, to express the merit function in terms of applied stress. For axially loaded uniform members this can always be done as follows:

The cross-sectional area can be written as $A = P/\sigma_A$ therefore

$$W = \rho \frac{P}{\sigma_A} L$$

or

$$W = \frac{PL}{\left(\dfrac{\sigma_A}{\rho}\right)} \tag{2-9}$$

and since the product PL is fixed by the environment, the weight will be minimum if the quantity σ_A/ρ is a maximum. We can, therefore, take as an equivalent merit function for axially loaded uniform members the requirement that

$$\frac{\sigma_A}{\rho} \longrightarrow \max \qquad (2\text{-}10)$$

In order to algebraically manipulate the Euler buckling constraint, the radius of gyration must be expressed as a function of the proportion variables as follows:

for a thin-walled tube $\begin{cases} A = \pi D t \\ I = \dfrac{\pi D^3 t}{8} \end{cases}$

therefore

$$r^2 = \frac{I}{A} = \frac{D^2}{8} \qquad (2\text{-}11)$$

Combining Eqs. (2-5) and (2-11) yields

$$\sigma_A = \psi_E \frac{\pi^2 \eta_T E D^2}{8 c^2 L^2} \qquad (2\text{-}12)$$

Noting that Eq. (2-7) contains (t/D) and that Eq. (2-12) contains D^2, we can rewrite Eq. (2-10) as

$$\frac{\sigma_A}{\rho} = \frac{P}{\rho A} = \frac{P}{\rho \pi D t} = \frac{P}{\rho \pi D^2}\left(\frac{D}{t}\right) \qquad (2\text{-}13)$$

D^2 can be written in terms of applied stress from Eq. (2-12)

$$D^2 = \frac{8 c^2 L^2 \sigma_A}{\psi_E \pi^2 \eta_T E} \qquad (2\text{-}14)$$

Combining Eqs. (2-13) and (2-14), and after some algebraic manipulation, we obtain

$$\frac{\sigma_A}{\rho} = \frac{\psi_E^{1/2}}{c}\left(\frac{\pi}{8}\right)^{1/2} \eta_T^{1/2} \frac{E^{1/2}}{\rho}\left(\frac{D}{t}\right)^{1/2}\left(\frac{P}{L^2}\right)^{1/2} \qquad (2\text{-}15)$$

The merit function, now in terms of the ratio D/t, can be be combined with the local buckling constraint by substituting for D/t its equivalence from Eq. (2-7) as follows:

$$\frac{D}{t} = \frac{\psi_L K_c \eta_T^{1/2} E}{\sigma_A} \qquad (2\text{-}16)$$

Combining Eqs. (2-15) and (2-16), and after some algebraic manipulation, we obtain

$$\frac{\sigma_A}{\rho} = \frac{\psi_E^{1/3} \psi_L^{1/3}}{c^{2/3}}\left(\frac{\pi}{8}\right)^{1/3} K_c^{1/3} \eta_T^{1/2}\left(\frac{E^{2/3}}{\rho}\right)\left(\frac{P}{L^2}\right)^{1/3} \qquad (2\text{-}17)$$

Evaluating for $K_c = 0.40$, and also insisting on satisfaction of the yield, Euler, and local constraints, results in

$$\frac{\sigma_A}{\rho} = \frac{0.54}{c^{2/3}} \psi_E^{1/3} \psi_L^{1/3} \eta_T^{1/2} \left(\frac{E^{2/3}}{\rho}\right) \left(\frac{P}{L^2}\right)^{1/3} \quad (2\text{-}18)$$

$$\sigma_A \leq \sigma_y$$
$$\psi_E \leq 1$$
$$\psi_L \leq 1$$

From Eq. (2-18), it can be seen that for any given material $E^{2/3}/\rho$ and environment P/L^2,* the merit function will be maximum when both ψ_E and ψ_L are maximum. Since under the constraints, ψ_E and ψ_L must be less than or equal to unity, we conclude that

$$\psi_{E_{\text{opt}}} = \psi_{L_{\text{opt}}} = 1 \quad (2\text{-}19)$$

which implies simultaneous occurrence of Euler and local buckling at the applied stress for minimum weight. Equation (2-18) thus becomes

$$\left(\frac{\sigma_A}{\rho}\right)_{\text{opt}} = \frac{0.54}{c^{2/3}} \eta_T^{1/2} \left(\frac{E^{2/3}}{\rho}\right) \left(\frac{P}{L^2}\right)^{1/3} \quad (2\text{-}20)$$

$$\frac{\sigma_A}{\rho} \leq \frac{\sigma_y}{\rho}$$

It can be seen that low values of load index correspond to low values of optimum applied stress; in which case the yield constraint is conservatively satisfied. In these (elastic) design regions, the measure of efficiency of a material is

$$\frac{E^{2/3}}{\rho} \longrightarrow \max \quad (2\text{-}21)$$

for elastic design (low P/L^2). For sufficiently high values of the load index, where Eq. (2-20) predicts $\sigma_A > \sigma_y$, the yield constraint overrides and the appropriate material metric becomes

$$\frac{\sigma_y}{\rho} \longrightarrow \max \quad (2\text{-}22)$$

for inelastic design (high P/L^2).

The specific value of P/L^2 for which the material metric changes from $E^{2/3}/\rho$ to σ_y/ρ is illustrated in Fig. 2-5, where Eq. (2-20) has been plotted for various materials. Note that Mg AZ61A is the optimum material in the elastic design range $(0 < P/L^2 < 2)$ where $E^{2/3}/\rho$ is the governing material metric and that Ti-6A1-4V is the optimum material in the inelastic design range $(P/L^2 > 35)$ where σ_y/ρ is the governing material metric. In the tran-

---

*Referred to respectively as the *material metric* and the *load index*.

sition design range (moderate P/L^2), the optimum material is seen to be Al 7075-T6.

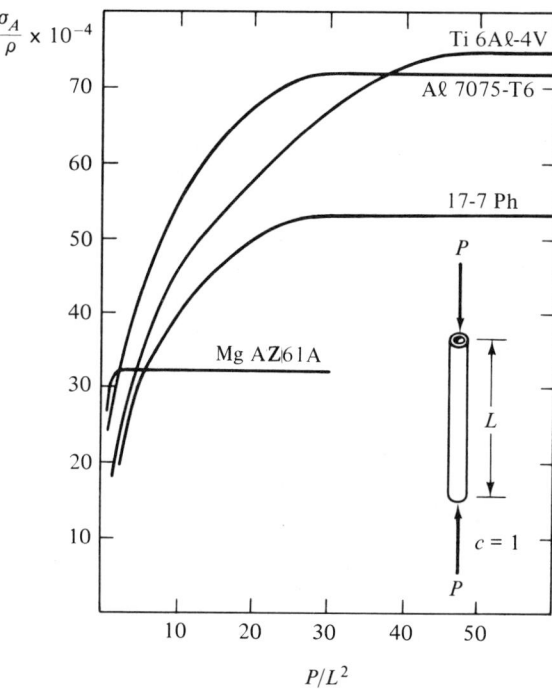

Fig. 2-5. Optimum Merit Function for a Circular Tube Column

Curves for selected materials of optimum applied stress ("weighted by the density factor") as a function of a load index represents a useful design chart in that it displays for a continuous expression of the environment the optimum stress and material for a specified form.

The optimum values of the proportion variables, D_{opt} and t_{opt}, can be evaluated by first employing the relationship of applied stress in terms of load and cross-sectional area as follows:

$$\frac{P}{\pi D_{\text{opt}} t_{\text{opt}}} = \sigma_{A_{\text{opt}}} \tag{2-23}$$

Secondly Eq. (2-14) can be employed with $\sigma_A = \sigma_{A_{\text{opt}}}$ and $\psi_{E_{\text{opt}}} = 1$ and yields

$$D_{\text{opt}}^2 = \frac{8c^2 L^2 \sigma_{A_{\text{opt}}}}{\pi^2 \eta_T E} \tag{2-24}$$

Note that although $\sigma_{A_{\text{opt}}}$ can be evaluated from a specification of the load index (the ratio P/L^2), the optimum values of D and t can only be evaluated from Eqs. (2-23) and (2-24) for specific values of P and L.

It should be noted that although for $\sigma_{A_{opt}} \leq \sigma_y$, ψ_E and ψ_L are unity for an optimum design, for a fully stressed design ($\sigma_{A_{opt}} = \sigma_y$) a range of values can be determined for both ψ_E and ψ_L which results in the same optimum weight. This is illustrated in Fig. 2-6. Suppose, as shown in the figure, that $\sigma_{A_{opt}}$ was sufficiently large so that the yield constraint was invoked. Note that the value of $\sigma_{A_{opt}}$ predicted by direct application of Eq. (2-20) without considering the yield constraint would result in an over stressed design, and consequently we must reduce this stress to σ_y. As a result, any selection of the product $(\psi_E \psi_L)^{1/3}$ which causes Eq. (2-18) to diminish to a value still greater than σ_y will be optimum. We see, therefore, that for a fully stressed design the SMD is only one of a family of optimum designs.

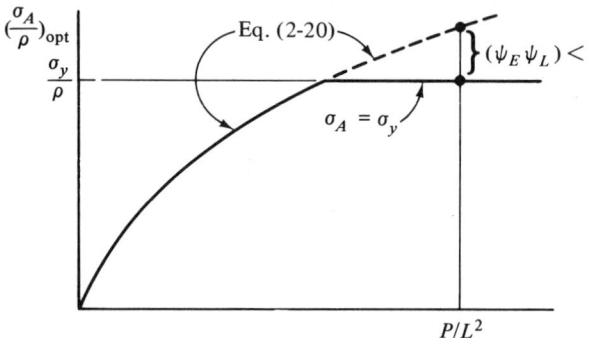

Fig. 2-6. Nonsimultaneous Mode Designs at the Fully Stressed Range

Section 3

THE *H*-SECTION COLUMN

Consider the cross section of Fig. 2-7.

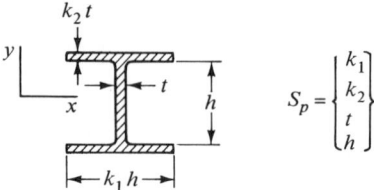

Fig. 2-7. The *H*-Section Column

The form has been defined in terms of four proportion variables as shown in the figure. The reason for introducing the ratios k_1, k_2 (as opposed to a flange thickness and flange width) will be seen upon subsequent mathematical evaluation, which will result in optimum values of k_1 and k_2 independent of a numerical expression of the environment, i.e., they are secondary system variables.

The Euler constraint, Eq. (2-5), will apply here, as well as the "applied stress" merit function, Eq. (2-10). We must also admit to the possibility of local buckling of the flange and web elements. These buckling modes are referred to as *plate buckling* and are characterized in an alternating pattern of buckled waves as shown in Fig. 2-8.

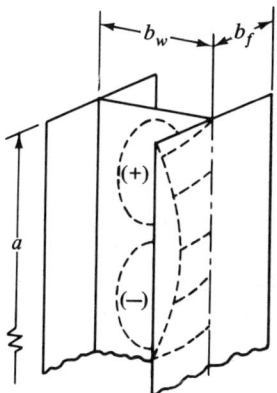

Fig. 2-8. Local Buckling of an *H*-Section

The predicting equation for buckling failure stress takes the form[11]

$$\sigma_L = K_p \eta_T^{1/2} E \left(\frac{t}{b}\right)^2 \qquad (2\text{-}25)$$

where t is the plate thickness and b is the plate width (normal to the direction of applied stress). K_p is a buckling coefficient, and depends on the conditions of edge support and "aspect ratio" (a/b) (a is the longitudinal plate dimension parallel to the direction of applied stress). For large aspect ratios ($a/b > 1$), K_p is a function of edge support only. Also, the buckled flange, with one longitudinal edge free, is likely to have a long half-wave length compared to its width. The web, on the other hand, being supported at both longitudinal edges, will buckle in a half-wave length comparable to its width. Accordingly, it is unreasonable to suppose a strong coupling between flange and web when buckling ensues. In the absence of this coupling an assumption of simple supports at the web-flange intersection will be reasonable.* With this assumption the buckling coefficients for flange and web become $K_p = 0.385$ and 3.62 respectively. Hence from Eq. (2-25)

$$\sigma_{L_f} = 0.385\eta_T^{1/2} E\left(\frac{k_2 t}{k_1 h/2}\right)^2 \qquad (2\text{-}26)$$

$$\sigma_{L_w} = 3.62\eta_T^{1/2} E\left(\frac{t}{h}\right)^2 \qquad (2\text{-}27)$$

It has been noted as an intuitive result that maximum efficiency will occur when the cross section is balanced in the Euler mode.† Equating I_{xx} to I_{yy} results in

$$\underbrace{\frac{th^3}{12} + 2(k_1 h)(k_2 t)\left(\frac{h}{2}\right)^2}_{I_{xx}} = \underbrace{\frac{2k_2 t(k_1 h)^3}{12}}_{I_{yy}}$$

from which

$$1 + 6k_1 k_2 = 2k_1^3 k_2 \qquad (2\text{-}28)$$

In the past, optimization procedures have dealt with situations involving more than one local buckling mode by employing the SMD philosophy. It is a tempting proposition to suppose that, at optimum conditions, flange and web buckle simultaneously. Under this assumption, equating Eqs. (2-26) and (2-27) results in a relation between the k's as shown below:

$$\sigma_{L_f} = \sigma_{L_w} \Longrightarrow k_2 = 1.53\, k_1 \qquad (2\text{-}29)$$

Note that Eqs. (2-28) and (2-29), when solved simultaneously, yield values for k_1 and k_2 (independent of load) and reduce the problem immediately to two primary system variables, t and h; obviously a desirable state of affairs. To equate any two modes (or any mode to the applied stress), however, may limit the scope of the solution (refer back to Example 1-2). Recognize that, in the H-section, the flanges are more effective as contributers to the moment of inertia than the web. It is therefore conceivable that at the optimum design the web buckles with flanges conservative. Although the SMD philosophy implies that a conservative flange represents extraneous material distribution, it must be emphasized that this implication only applies to the local buckling characteristics of the cross section. Since a conservative flange will possibly enhance the resistance to general (Euler) buckling, the question of overall

---

*In the presence of coupling the condition of support will be somewhere between free and simple supports.

†This can be demonstrated rigorously by evaluating for the two separate conditions, $I_{xx} \geq I_{yy}$ and $I_{xx} \leq I_{yy}$. In each case, mathematical optimization results in the equality as optimum.

efficiency can only be settled by analytical evaluation. We can consider the general problem of a possibly conservative flange at optimum weight by stipulating the following condition

$$\sigma_{L_w} \leq \sigma_{L_f} \tag{2-30}$$

Combining Eqs. (2-26), (2-27), and (2-30) yields

$$k_2 \geq 1.53 \, k_1 \tag{2-31}$$

Note that we have not ruled out the possibility of a SMD optimum, which appears as the equality of Eq. (2-30) or (2-31).

We are now in a position to combine the failure constraints with the merit function. The cross-sectional area for the H-section can be written as

$$A = 2k_2 t k_1 h + th = th(1 + 2k_1 k_2) \tag{2-32}$$

Combining Eq. (2-32) with Eq. (2-10) yields the merit function in terms of t, h, k_1 and k_2 as follows:

$$\frac{\sigma_A}{\rho} = \frac{P}{\rho A} = \frac{P}{\rho th(1 + 2k_1 k_2)} \tag{2-33}$$

To convert the Euler constraint to a function of t and h we can evaluate the radius of gyration as

$$r^2 = \frac{I_{yy}}{A} = \frac{k_1^3 k_2}{6(1 + 2k_1 k_2)} h^2 \tag{2-34}$$

Now, combining Eqs. (2-34) and (2-5) yields

$$\sigma_A = \psi_E \frac{\pi^2 \eta_T E k_1^3 k_2}{6c^2 L^2 (1 + 2k_1 k_2)} h^2 \tag{2-35}$$

$$\psi_E \leq 1$$

Selecting Eq. (2-27) as our local constraint* and introducing a slack variable results in

$$\sigma_A = \psi_L 3.62 \eta_T^{1/2} E \left(\frac{t}{h}\right)^2 \tag{2-36}$$

$$\psi_L \leq 1$$

Noting that Eq. (2-35) contains h^2 and that Eq. (2-36) contains t/h, we can rewrite the merit function, Eq. (2-33), as follows:

$$\frac{\sigma_A}{\rho} = \frac{P}{\rho(1 + 2k_1 k_2)\left(\frac{t}{h}\right) h^2} \tag{2-37}$$

---

*Since web will be more critical than flange, i.e., $\sigma_{L_w} \leq \sigma_{L_f}$.

Eliminating h^2 and t/h from Eq. (2-37) by combining with Eqs. (2-35) and (2-36) yields

$$\frac{\sigma_A}{\rho} = \frac{E^{3/5}}{\rho}\left[\frac{k_1^3 k_2}{(1+2k_1 k_2)^2}\right]^{2/5} \frac{(3.62)^{1/5}\pi^{4/5}}{(6)^{2/5}c^{4/5}} \psi_L^{1/5}\psi_E^{2/5}\eta_T^{1/2}\left(\frac{P}{L^2}\right)^{2/5} \quad (2\text{-}38)$$

subject to

$$\sigma_A \leq \sigma_y$$
$$\psi_L \leq 1, \psi_E \leq 1$$
$$(1+6k_1 k_2) = 2k_1^3 k_2 \quad (2\text{-}39)$$

and

$$k_2 \geq 1.53 k_1$$

We can see immediately that $\psi_{L_{opt}} = \psi_{E_{opt}} = 1$, and also that the following $f_e(k_1, k_2)^*$ must be maximized subject to the constraints of Eq. (2-39)

$$f_e(k_1, k_2) = \left[\frac{(3.62)^{1/2}\pi^2 k_1^3 k_2}{6(1+2k_1 k_2)^2}\right]^{2/5} \longrightarrow \max \quad (2\text{-}40)$$

subject to

$$k_2 \geq 1.53 k_1$$

and

$$1 + 6k_1 k_2 = 2k_1^3 k_2 \quad \text{or} \quad k_2 = \frac{1}{2k_1^3 - 6k_1} \quad (2\text{-}41)$$

The constraints of Eq. (2-41) are shown below in Fig. 2-9.

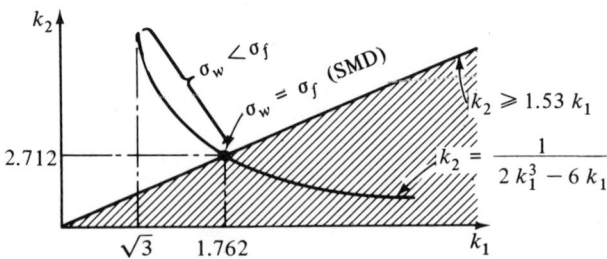

Fig. 2-9. Permissible Values of k_1, k_2

It is apparent from Fig. 2-9 that permissible values of k_1 be between $\sqrt{3}$ and 1.762 where the corresponding values of k_2 are obtained from Eq. (2-41). Fig. 2-10 shows $f_e(k_1, k_2)$ plotted as a function of k_1 and concludes that a conservative flange represents a weight penalty over the SMD.

*Referred to as the "efficiency function."

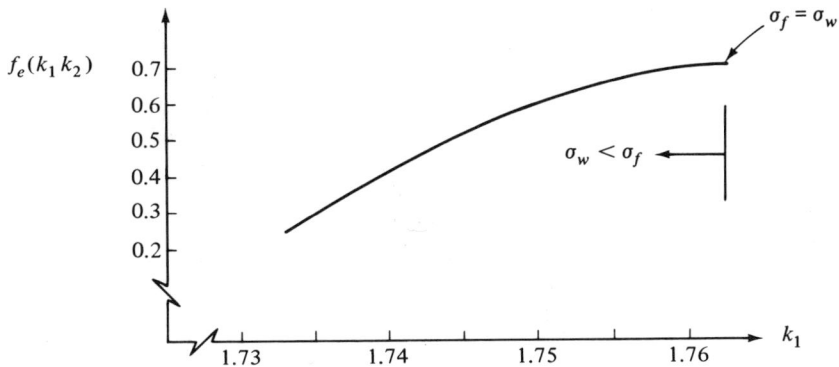

Fig. 2-10. $f_e(k_1, k_2)$ versus k_1

Accordingly $f_{e_{max}} = 0.705$ and Eq. (2-38) becomes

$$\left(\frac{\sigma_A}{\rho}\right)_{opt} = \frac{0.705}{c^{4/5}} \eta_T^{1/2} \frac{E^{3/5}}{\rho} \left(\frac{P}{L^2}\right)^{2/5} \quad (2\text{-}42)$$

$$\sigma_A \leq \sigma_y$$
$$k_{1\,opt} = 1.762$$
$$k_{2\,opt} = 2.712$$

For a given design problem t and h can be evaluated from Eqs. (2-36) and (2-37). The material metric for elastic design is seen to be $E^{3/5}/\rho$. The reader should compare this result to that for the circular tube. Note that the load index is again P/L^2, a result inherent to the column environment. Fig. 2-11 shows Eq. (2-42) plotted for various materials. In Fig. 2-12 a graphical comparison is shown between the circular tube, square tube (see Prob. 2-3) and H-section column for an aluminum alloy.

Section 4
WEIGHT INDEX VERSUS LOAD INDEX

We have seen examples of an applied stress equivalent merit function evaluated in terms of material metric and load index. Oftentimes an actual weight evaluation is desired in these terms. These instances arise where comparisons are desired against forms for which the applied stress merit is not applicable. Furthermore, in the design of structural systems for which a column is a component element, a weight merit is required in order to be additive to the total system weight merit function.

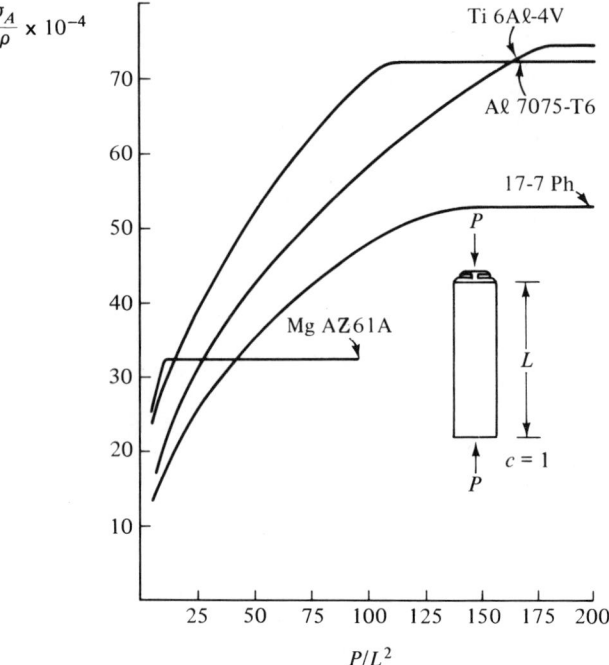

Fig. 2-11. Optimum Merit Function for an H-Section Column

For the axial force transmission members considered in this chapter the conversion is accomplished simply by introducing the appropriate expression for $(\sigma_A/\rho)_{opt}$ into Eq. (2-9). Recognizing that for elastic design $(\sigma_A/\rho)_{opt}$ is a function of P/L^2 and that for plastic design $(\sigma_A/\rho)_{opt} = \sigma_y/\rho$, we first convert Eq. (2-9) to P/L^2 index form by dividing through by L^3, which yields

$$\left(\frac{W}{L^3}\right)_{opt} = \frac{\left(\frac{P}{L^2}\right)}{\left(\frac{\sigma_A}{\rho}\right)_{opt}} \tag{2-43}$$

and thus identifies W/L^3 as a convenient weight index.

By introducing Eqs. (2-20) and (2-42) into Eq. (2-43) we find:
For the circular tube column

$$\left(\frac{W}{L^3}\right)_{opt} = \frac{c^{2/3}(P/L^2)^{2/3}}{0.54\eta_T^{1/2}(E^{2/3}/\rho)}, \quad \sigma_A < \sigma_y \tag{2-44}$$

$$\left(\frac{W}{L^3}\right)_{opt} = \frac{(P/L^2)}{(\sigma_y/\rho)}, \quad \sigma_A = \sigma_y \tag{2-45}$$

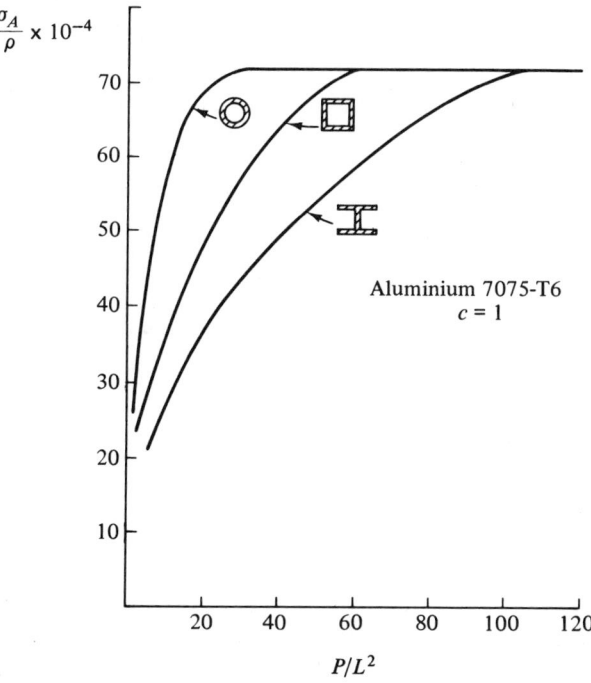

Fig. 2-12. Column Form Comparison for an Aluminum Alloy

For the H-section column

$$\left(\frac{W}{L^3}\right)_{\text{opt}} = \frac{c^{4/5}(P/L^2)^{3/5}}{0.705\eta_T^{1/2}(E^{3/5}/\rho)}, \quad \sigma_A < \sigma_y \tag{2-46}$$

$$\left(\frac{W}{L^3}\right)_{\text{opt}} = \frac{(P/L^2)}{(\sigma_y/\rho)}, \quad \sigma_A = \sigma_y \tag{2-47}$$

Note that in the plastic design range ($\sigma_A = \sigma_y$) the relationship W/L^3 as a function of P/L^2 is a straight line with slope ρ/σ_y, which passes through the origin (if extrapolated). Fig. 2-13 illustrates the form comparisons of Fig. 2-12 on the basis of weight index.

Section 5

CONDITIONS OF END SUPPORT

The optimum merit evaluations are applicable for general end supports, where c represents the ratio of the equivalent length (L_e) of the buckled half-wave to the actual length of the column. The following figure gives the values of c corresponding to various end conditions.

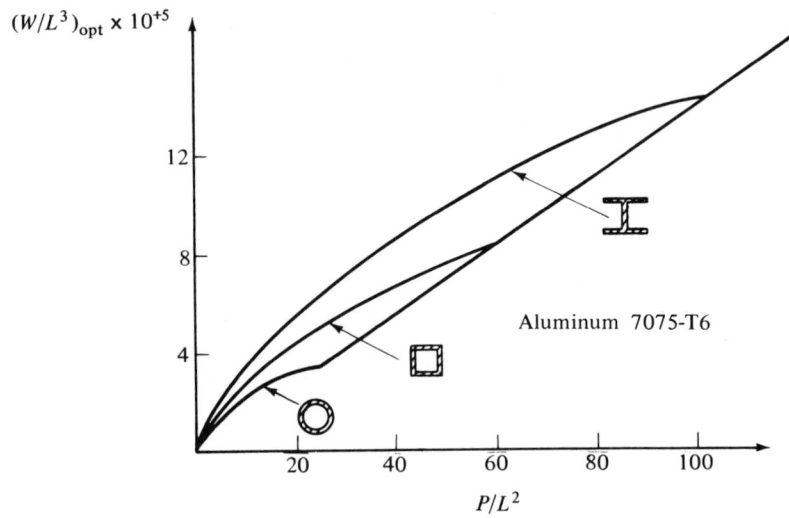

Fig. 2-13. Form Comparison of Fig. 2-12 on a Weight Basis

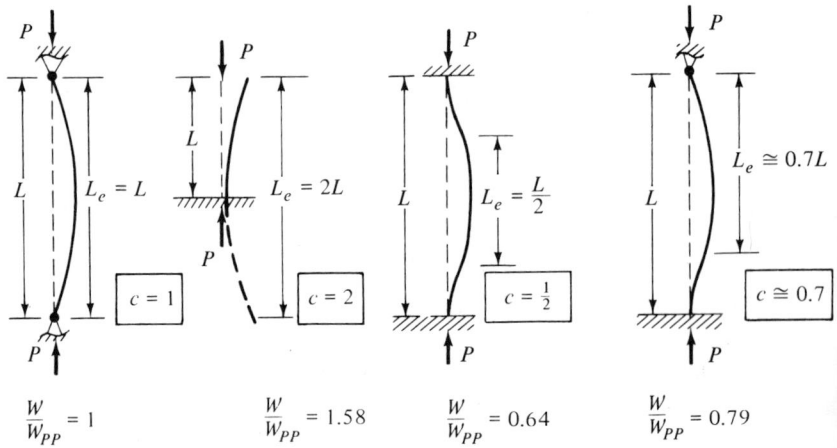

Fig. 2-14. Equivalent Lengths For Typical End Conditions

Shown below each case are the respective weight ratios for a circular tube, referenced to the pinned-pinned condition. From this, it can be seen that the fixed-fixed condition represents the most efficient manner of end support, as should be expected. Less obvious is the fact that, for the circular tube, the fixed-fixed condition represents a 36 percent weight savings over the pinned-pinned condition, a sizable savings which should, in many cases (moderately long and long columns), justify the fastening weight increment necessary to create this end condition.

Section 6
GEOMETRIC CONSTRAINTS

At the lower values of load index the open-variable optimum column is likely to violate a geometric constraint which prescribes a minimum thickness dimension. As an example of how slack variables can be optimized while satisfying such a constraint consider the circular tube column subject to a minimum thickness requirement $t \geq t_m$.

Here the number of intersecting failure modes is equal to the number of proportion variables since at the lower values of P/L^2 the yield constraint is conservatively suppressed. Accordingly it is possible to evaluate expressions for both thickness and merit in terms of the slack variables alone. Employing Eq. (2-7) and substituting $\sigma_A = P/(\pi D t)$ yields for t

$$t = \frac{P^{1/2}}{\pi^{1/2} K_c^{1/2} \eta_T^{1/4} E^{1/2} \psi_L^{1/2}} \tag{2-48}$$

Now consider simultaneously the expression for merit (Eq. (2-18)) reproduced below for convenience:

$$\frac{\sigma_A}{\rho} = \frac{0.54}{c^{2/3}} \psi_E^{1/3} \psi_L^{1/3} \eta_T^{1/2} \left(\frac{E^{2/3}}{\rho}\right) \left(\frac{P}{L^2}\right)^{1/3} \tag{2-49}$$

Suppose the value of t at the open variable optimum value of $\psi_L = 1$ is less than t_m. Considering the permissible values of the slack variables it is possible to effect the constraint by selecting the constrained optimum values of ψ_L and ψ_E as

$$\psi_{L\text{co}} = \left(\frac{t_{\text{opt}}}{t_m}\right)^2, \quad \psi_{E\text{co}} = 1 \tag{2-50}$$

Since the applied stress merit is proportional to $\psi_L^{1/3}$ the weight ratio, constrained optimum to open variable optimum, can be expressed as

$$\frac{W_{\text{co}}}{W_{\text{opt}}} = \frac{\psi_{L\text{opt}}^{1/3}}{\psi_{L\text{co}}^{1/3}} = \left(\frac{t_m}{t_{\text{opt}}}\right)^{2/3} \tag{2-51}$$

For example if $t_m = 2 t_{\text{opt}}$ the weight does not double but equals $2^{2/3}$. This suggests that a *redistribution of material* is effected which causes a reduction in the diameter as the thickness is increased.* To establish the corresponding constrained optimum value of D we can manipulate Eq. (2-12) with $\sigma_A = P/(\pi D t)$ and setting $t = t_m$ and $\psi_{E\text{co}} = 1$ yields

$$D_{\text{co}} = \frac{2 P^{1/3} L^{2/3} c^{2/3}}{\pi E^{1/3} \eta_T^{1/3} t_m^{1/3}} \tag{2-52}$$

The result is that the increased thickness constraint causes a conservative margin with respect to local buckling and a corresponding reduction in diameter at the onset of Euler buckling as given by Eq. (2-52).

*As discussed in Sec. 1-10.

PROBLEMS

2-1. A pinned circular tube column is to be designed for a length of 100 in. Using either aluminum or titanium, find optimum stress and dimensions for a load of:
(a) 100,000 lb (F.S. included)
(b) 500,000 lb (F.S. included)

2-2. A 7075-T6 aluminum circular tube is to be designed for an environment of $E_S = \{70 \text{ in}, 5000 \text{ lb}\}$ as a pinned column. Due to manufacturing feasibility the tube thickness is subject to the constraint $t \geq 0.03$ in. Find the optimum diameter using a graphical design space and compare with the analytical constraint solution.

2-3. For the square tube cross section shown below, show that the minimum weight optimum design equations are:

$$\left(\frac{\sigma_A}{\rho}\right)_{opt} = \frac{0.910}{c^{4/5}} \eta_T^{1/2} \frac{E^{3/5}}{\rho} \left(\frac{P}{L^2}\right)^{2/5}$$

$$\left(\frac{b}{t}\right)_{opt} = \frac{1.90 \eta_T^{1/4} E^{1/2}}{\sigma_{A_{opt}}^{1/2}}$$

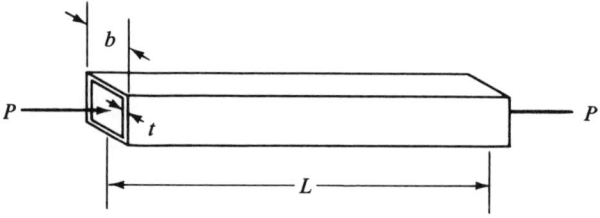

2-4. Compare the weight ratio of the square tube to the circular tube for the material and environment of Problem 2-1(a) and 2-1(b). What generalizations can be made about these comparisons?

2-5. Comment on the weight comparison between the square tube and the H-section column for the general environment.

2-6. As opposed to carrying P pounds with a single optimized circular tube it is suggested to carry $P/2$ pounds each with two identical optimized circular tubes as shown below.

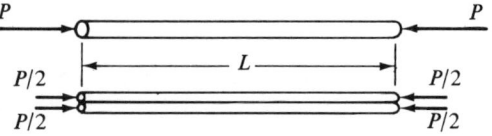

Show that this must always result in a weight penalty when $\sigma_{A_{opt}} < \sigma_y$.

2-7. Metallurgist A proposes a research effort on a metal whereby the modulus of elasticity will be increased by 10 percent. Metallurgist B proposes a research effort whereby the density will be reduced by 10 percent. Which effort would you fund?

2-8. A circular tube column is to carry a load of 500,000 lb over a pin-supported length of 100 in. From Fig. 2-5 the titanium design is clearly optimum at this load index. It is proposed, however, to split the load using two aluminum 7075-T6 tubes thereby taking advantage of this material's superiority over titanium at the lower value of index. Does this result in a weight savings?

2-9. Using the results of Prob. 2-3, show that an optimum weight merit function for the square tube column can be expressed as

$$\left(\frac{W}{L^3}\right)_{opt} = \rho \frac{1.10 c^{4/5}}{\eta_T^{1/2} E^{3/5}} \left(\frac{P}{L^2}\right)^{3/5}, \quad \sigma_A < \sigma_y$$

$$\left(\frac{W}{L^3}\right)_{opt} = \frac{\rho}{\sigma_y}\left(\frac{P}{L^2}\right), \quad \sigma_A = \sigma_y$$

2-10. Compute the percentage weight savings for fixed ends compared to pinned ends in an *H*-section column.

2-11. Develop Eq. (2-44) employing a weight merit function as opposed to an applied stress merit.

3
DESIGN BASED ON A COST-WEIGHT TRADE-OFF

Within conventional structural design there are problems which obviously fall into either strict cost minimization or strict weight minimization categories. For example, aside from the obvious limitation of failure due to the structure's own weight, a freeway off-ramp would be designed as a minimum cost system without regard to its weight.* At the other end of the spectrum we find the upper stages of rocket vehicles, and many of the critically loaded structural components of advanced aircraft subject to strict weight minimization. This is not to say that in these instances cost is totally disregarded. In any engineering design, the cost is bounded by some funded maximum. What is meant is that there are occasions where the successful performance of a vehicle hinges on the ability to design given structural components at or near their absolute minimum in weight. Consequently we find large sums of money allocated for the engineering design, fabrication, testing, and production of these components. Therefore, in these cases, cost can be disregarded (provisionally) and the merit function can be considered to be weight.

It has already been noted in Chap. 1, Sec. 5, that an attempt to simultaneously minimize weight and cost is generally contradictory. Typically the minimum weight configuration is the most expensive of the candidates. In many applications both the cost of the minimum weight structure and the weight of the minimum cost structure are prohibitive, resulting in a mandatory *trade-off* compromise wherein neither cost nor weight is the governing merit function.

Those problems for which neither cost nor weight can be singled out as the criterion for design are the concern of this chapter. Our objective will be to develop a merit function based on a cost-weight trade-off, where a *value comparator* has been established of the form: "dollars we are willing to

---

*There would obviously be regard to aesthetics in such a design besides the principal criterion of cost.

spend for each pound of weight saved."* Establishment of such a comparator enables the determination of a trade-off merit function as we shall presently see. Before developing a trade-off merit function on this basis, however, we will first consider the manner in which a dollar value of a pound comparator is established.

Section 1
THE DOLLAR VALUE OF A POUND

How does one determine the dollar worth of saving a pound of weight? Consider first the case of a mobile structure (vehicle). A vehicle is generally designed for a given performance requirement over some operational life. Even prior to designing the load-carrying structure, an actual expenditure in terms of dollars for each pound of structural weight can be assessed based on fuel and engine requirements to carry each pound over the operational life of the vehicle. This alone establishes a dollar value of saving weight and can be used as a component of the value comparator for subsequent design of the load-carrying structure. An additional component arises due to the desirous effect of reduced structural weight on performance which, if assessable, may yield additional potential dollar worth for each pound of weight saved.

Certainly, in the above evaluations, adjustments must be made due to the economic effect of spending a dollar now on structure as opposed to the equivalent, but gradual, expenditure in fuel over the operational life. There is also the implication of an iterative approach to establishing the dollar worth of saving weight with respect to the effects on vehicle performance. In such instances, after a preliminary weight estimation the dollar worth of saving weight is established, and is applied in a redesigning of the structure.

For an immobile structure (civil engineering type structure) the component of the value comparator dealing with weight's effect on operational cost is nil. There can, however, be a dollar worth of saving weight assessed with respect to the "performance" of such a structure. Excessive weight structures in these applications, even though optimum from the standpoint of cost, may be prohibitive when it comes to meeting performance requirements. Examples of such performance requirements would be limitations on the inertial aspect of excessive weight (which might be undesirable in a dynamic response due to earthquakes), restriction of the number of stories in a dwelling due to excessive accumulation of weight. In the latter case a dollar worth of saving weight simply evolves from an assessment of the projected profits associated with an increased building capacity. In the former case the establishment of dollar worth is by nature a qualitative judgment since here the

*We shall denote this value comparator by V (in units of $/lb).

dollars paid per pound must somehow be equated to the increased safety and reliability associated with the reduced weight.

We see, therefore, that the dollar worth of saving a pound of weight ($V\$/lb$) is at least a potential outcome in the establishment of the criteria phase of design. It should also be apparent that without such a value comparator, a cost-weight trade-off must proceed from an ill-defined base and is qualitative at best.

Section 2

EVALUATION OF WEIGHT AND COST MERIT FUNCTIONS

By definition a merit function must be configuration dependent and rank candidates such that a stationary value implies an optimum condition. We have already seen that the weight of a structure can be expressed analytically in terms of configuration (as a function of proportions for a given form). An analytical cost merit function which predicts cost as a function of proportions is a more difficult proposition. In a trade-off study for which a compromise is sought between the low-weight/high-cost and high-weight/low-cost candidates, the cost is obviously not just material expense (if it were, low weight and low cost would be coincident goals). The property of the minimum weight configuration which results in high cost has to do with the "exotic" nature of light weight concepts. In these cases the principal part of the cost merit function results from machining and fabrication processes in addition to cost of material. Machining and fabrication entail equipment and labor costs and must, in general, be evaluated for each specific candidate independently.

In some cases, empirical expressions are evolved which predict these costs as a function of those proportion variables which critically effect cost. As an example, consider the case of an integrally stiffened plate cross section as shown in Fig. 3-1.

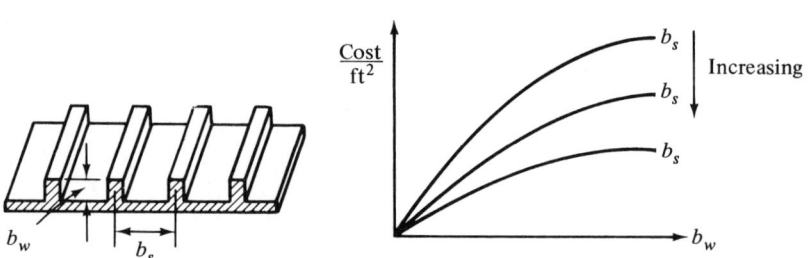

Fig. 3-1. Machining Cost as a Function of Proportions

The machining cost per square foot of plate will depend on the depth of mill (b_w) between stiffeners, and there will also be a cost reduction assessable with respect to increasing distance between stiffeners (b_s) since, as this dimension increases, there will be fewer stiffener lines, each of which requires machining to tolerance.

In the absence of an empirical expression for the cost merit function, the costs for a discrete number of candidates must be determined. Although this does not result in an analytical expression, it does yield a tabular merit function of cost in terms of proportions.*

Section 3

THE COST-WEIGHT TRADE-OFF MERIT FUNCTION

Consider a structural design problem for which the means exist for evaluating both the weight and cost of alternatives as a function of configuration. Assume that dollar worth of saving a pound of weight, $V\$/\text{lb}$, has been established. Suppose further that alternative configurations under consideration do not admit to an acceptable candidate which simultaneously minimizes both weight and cost.

We desire a trade-off compromise which yields a candidate weight for which those candidates with lower weights require spending more than $V\$/\text{lb}$ for each pound of reduced weight. Accordingly we make the following definition for a cost-weight trade-off merit function:

> The cost-weight trade-off merit function, M_{cw}, is a configuration-dependent function which attains a minimum for that acceptable candidate, such that an attempt to further reduce its weight requires an expenditure of more than V dollars per pound of reduced weight. (Note that this definition is consistant with that given generally for an optimum configuration in Chap. 1, Sec. 5.)

The determination of M_{cw} can be accomplished as follows. Imagine a cost-weight space for which cost is the ordinate and weight is the abscissa. Since both cost and weight can be evaluated for each acceptable candidate, all such configurations can be represented as points in such a space. An illustration of cost-weight space is shown in Fig. 3-2 on p. 56.

Consider any two points in cost-weight space, referred to as the ith and jth candidates. Assume the property that for any two alternative candidates the higher cost candidate has a lower weight.† Take the ith candidate as

---

*Such a table may be established between alternative forms in addition to the acceptable candidates of a given form. It will be shown that a trade-off is possible for both.

†Otherwise there can be no trade-off. There is no compromise required for a candidate with both least weight and least cost.

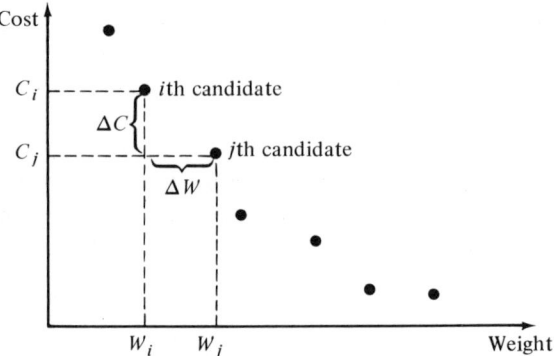

Fig. 3-2. Cost–Weight Space

having the higher cost and lower weight compared to the jth candidate. According to our definition for M_{cw}, the ith candidate can be ranked better than the jth if less than V dollars must be expended for each pound of weight difference between the two alternatives. Mathematically this yields the following condition:

$$\frac{\text{Cost (increase)}}{\text{Weight (reduction)}} = \frac{C_i - C_j}{W_j - W_i} < V \$/\text{lb} \tag{3-1}$$

Collecting like terms in Eq. (3-1) and rearranging results in

$$W_i + \frac{C_i}{V} < W_j + \frac{C_j}{V} \tag{3-2}$$

We see, therefore, that for a cost-weight relative optimum the sum $W + C/V$ must be less than that for the alternative candidate. Obviously the candidate which yields an absolute minimum for $W + C/V$ is the absolute optimum based on the trade-off criterion. The cost-weight trade-off merit function is thus found to be*

$$M_{cw} = W + \frac{C}{V} \longrightarrow \min \tag{3-3}$$

Note that in a case where the minimum weight configuration is also the minimum cost configuration, Eq. (3-3) will be minimum at this point in cost-weight space independent of V. Furthermore, as V grows large compared to the relative costs of alternatives, $M_{cw} \to \min$ reduces to the expected result of $W \to \min$. For the other extreme of the dollar worth of saving weight approaching zero, the term C/V overpowers and the expression

---

*First developed by Emero and Alvey[1,2] using a graphical approach based on the "optimum slope" in cost-weight space (see Prob. 3-5).

$M_{cw} \to$ min reduces to strict cost minimization. These trends are illustrated in Fig. 3-3.

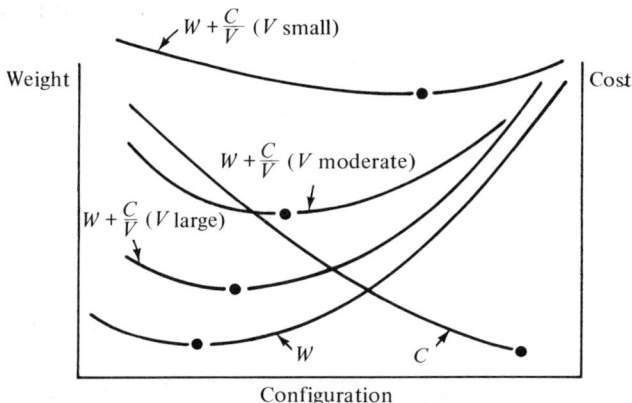

Fig. 3-3. Effect of V on the Trade-off Optimum

Note in the figure that for large V the trade-off merit curve predicts a minimum close to the minimum weight configuration and that for small V the trade-off minimum approaches the cost minimum. For some intermediate value of V the merit curve is shown to predict an optimum which represents a trade-off compromise.

Section 4
TRADE-OFF OPTIMIZATION OF PROPORTIONS FOR A GIVEN FORM

To illustrate the application of Eq. (3-3) in the design of conventional structures consider the following problem. A zee-stiffened wide column panel is specified for a load environment of a distributed axial load q(lb/in) over a transmission path L as shown in Fig. 3-4 on p. 58.

We shall consider the weight optimization of alternative forms for the general wide column environment (q, L) in Chap. 7. Suffice it to say here that the weight merit function can be optimized employing the technique of analytical optimization, yielding the set $\{S_p\}_{\text{opt}}$ for minimum weight. For typical numerical values of q and L these values are such that the cost of the minimum weight cross section may be prohibitive. In the present case prohibitive cost results primarily from the relatively close spacing of zee-stiffeners at optimum weight. For example, using aluminum 7075-T6 for $q = 8000$ lb/in and $L = 20$ in, b_s is found to be only 1.54 in for minimum weight. Aside from the

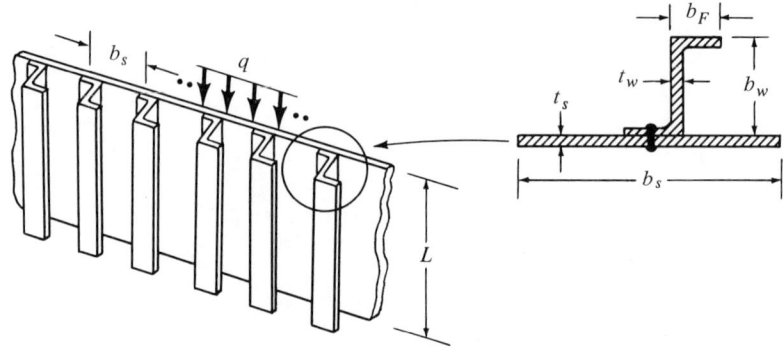

Fig. 3-4. The Zee-stiffened Wide Column

problems of associated fastening weight and possible violation of minimum gauge requirements, the cost of producing such a cross section would be high due to the large number of pieces and lines of attachment requiring fastening.

If, in a given design study, b_s is increased from its weight optimum value, the increased weight can be evaluated analytically from the weight merit function under this geometric constraint. The associated reduced cost can be assessed for values of b_s and thus a table displaying both weight and cost for specific values of b_s can be formulated. Such a study was undertaken at North American Aviation, Los Angeles Division.[12] Table 3-1 is the result of that study for a zee-stiffened wide column of Al 7075-T6, $q = 8000$ lb/in and $L = 20$ in.* The entries on the far right represent the evaluation of the trade-off merit function based on a value comparator of $V = 100$ \$/lb.

Table 3-1 Weight, Cost, and Cost-weight Trade-off in Terms of Stiffener Spacing for a Zee-Stiffened Wide Column

b_s(in)	$q = 8000$ lb/in, $L = 20$ in		$V = 100$ \$/lb	
	W(lb/ft²)	C(\$/ft²)	$(W + C/V)$ (lb/ft²)	
1.54	5.23	87.5	6.10	W_{opt}
1.76	5.25	83.8	6.09	
2.05	5.28	78.9	6.07	$M_{cw_{opt}}$
2.35	5.33	75.2	6.08	
2.77	5.58	72.9	6.31	
3.22	5.97	71.2	6.68	C_{opt}

*By permission of the American Institute of Aeronautics and Astronautics.

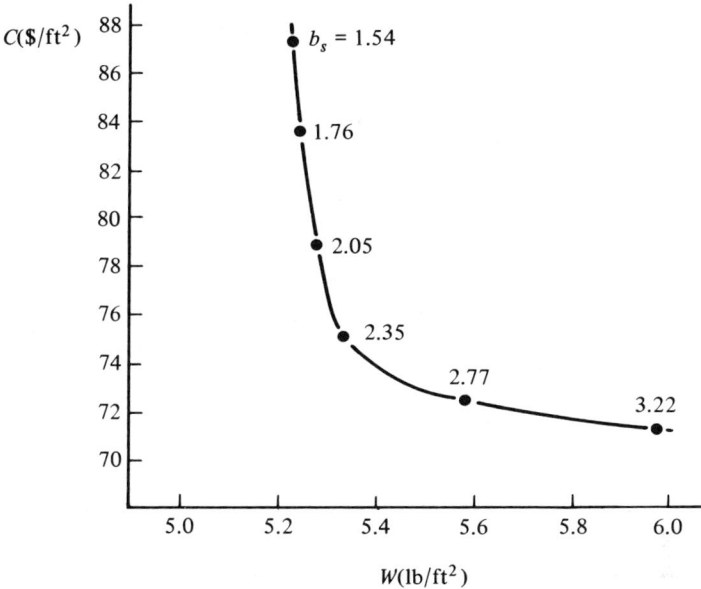

Fig. 3-5. Cost–weight Space, Zee-stiffened Wide Column for Aluminum 7075 T-6 at $q = 8000$ lb/in and $L = 20$ in

For a minimum in $(W + C/V)$, $b_s = 2.05$ in and represents an optimum compromise between the extremes of excessive cost and excessive weight. As a check on this conclusion note from the table that to reduce b_s from 2.05 in to 1.76 in results in a cost increase of 4.9 \$/ft² and a weight reduction of 0.03 lb/ft², the ratio of which equals 163 \$/lb and hence exceeds the established value comparator of 100 \$/lb.

The associated cost-weight space for this example is shown in Fig. 3-5. In Fig. 3-6 a graphical evaluation of $(W + C/V)$ is shown.* Note that using graphical interpolation the minimum point on the trade-off merit curve is found to be $b_s = 1.95$ in as opposed to the tabular evaluation of 2.05 in. This refinement is only justified if the associated cost-weight space is fairly continuous as it is in this case. In cases where the cost evaluation involves large jump discontinuities, the tabular evaluation cannot be interpolated. It must be emphasized that since b_s has been determined to be the only proportion variable which critically effects cost, the determination of the remaining proportion variables should be with respect to minimum weight. The preceding evaluation, therefore, presupposes that the weight merit can be optimized for given geometric constraints on b_s (as implied in Table 3-1 and Figs. 3-5 and 3-6). We shall consider the problem of weight optimization under geo-

*By permission of the American Institute of Aeronautics and Astronautics.

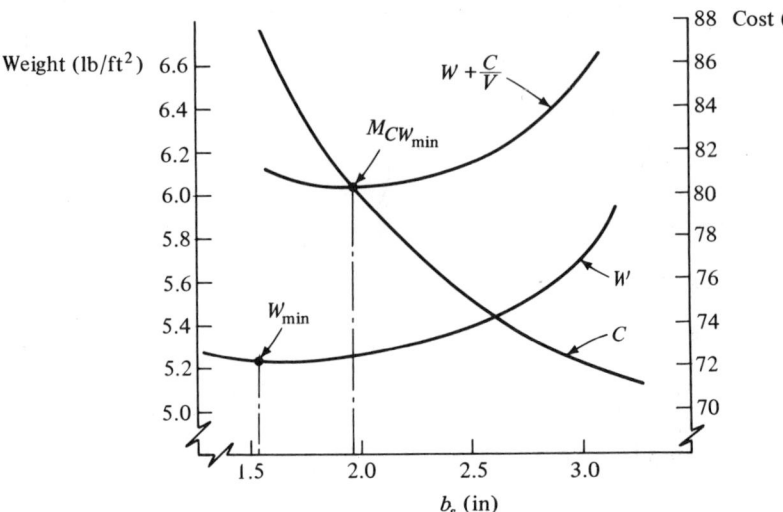

Fig. 3-6. Cost–weight Trade-off Optimization of b_s for $V = 100$ $/lb

metric constraints for wide columns in Chap. 7. We note here the importance of such considerations in a cost-weight trade-off evaluation.

Section 5
TRADE-OFF OPTIMIZATION BETWEEN ALTERNATIVE FORMS

As in the case of alternative candidates of a given form, an analytical trade-off determination between two forms is only necessary if the higher weight alternative has the lower cost. A determination of the trade-off optimum simply follows from the discrete evaluations of $W + C/V$ for each form.

EXAMPLE 3-1. As an alternative to the zee-stiffened trade-off optimum of the previous section, an integrally stiffened cross section is proposed as shown in Fig. 3-7. Assume that for the same environment and V, a trade-off optimum has been determined for the integrally stiffened cross section to be $W = 5.47$ lb/ft² and $C = 57.0$ $/ft². Determine if this represents a more efficient trade-off design than the zee-stiffened alternative.

We first recognize the need for a trade-off merit evaluation since the integral cross section represents increased weight at reduced cost. (See Table 3-1 on p. 58.)

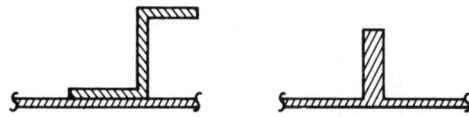

Fig. 3-7. Zee-stiffened and Integrally Stiffened Alternatives

We know from the previous section that, for the zee-stiffened cross section, $W + C/V = 6.07$. For the integral cross section we find that

$$W + \frac{C}{V} = 5.47\frac{\text{lb}}{\text{ft}^2} + \frac{57\$/\text{ft}^2}{100\$/\text{lb}} = 6.04\frac{\text{lb}}{\text{ft}^2} \quad (3\text{-}4)$$

and, therefore, the integrally stiffened cross section is shown to be more efficient based on a cost-weight trade-off.

Recognize that the above evaluation could be effected directly from Eq. (3-1) thus:

$$\frac{C_i - C_j}{W_j - W_i} = \frac{\$78.9 - \$57.0}{5.47\text{lb} - 5.28\text{lb}} = 115\$/\text{lb} \quad (3\text{-}5)$$

Since this exceeds the value comparator of 100 $/lb the lower cost alternative (the integral cross section in this case) is found to be more efficient as before. This approach, however, may be more desirable for a trade-off between two alternatives since it indicates the amount of dollar worth increment that would be required to swing the trade-off over to the alternative. The utility of the merit function approach (Eq. 3-3) is realized when a large number of alternatives must be compared and is essential in a continuous variable evaluation as illustrated in the previous section. Recognize that once a merit evaluation has been achieved neighboring designs can then be compared on the basis of Eq. (3-1). Note that this was done following Table 3-1.

PROBLEMS

3-1. Using Table 3-1 determine the value of the dollar worth of saving a pound of weight (V) such that the minimum cost configuration is also the trade-off optimum.

3-2. Using Table 3-1 determine the value of V such that the minimum weight configuration is also the trade-off optimum.

3-3. For some hypothetical load-carrying function an aluminum unstiffened plate is optimized for weight at a thickness of 0.02 in. A production cost assessment results in the fact that any gauge below a minimum of 0.05 in requires special tolerance control at an increased cost of 20 $/ft² of plate.

Determine the minimum value of V such that the lower gauge plate is the trade-off optimum.

ANSWER $V = 46.3$ \$/lb

3-4. Demonstrate that the point in a continuous cost-weight space which is the trade-off optimum has a tangent of slope $-V$.

3-5. Using the result of Prob. 3-4 derive the expression for trade-off merit, Eq. (3-3).

3-6. Determine the trade-off optimum for the example of Sec. 3-4 using a value comparator of $V = \$50/\text{lb}$.

3-7. A tensile member is to carry 10,000 lb over a transmission length of 70 in. Of the two materials, titanium 6Al-4V and AISI 1025 steel, which is the trade-off optimum based on a value comparator of $V = 3$ \$/lb? Assume the respective unit weight costs as follows: Ti, 8 \$/lb; steel, 0.08 \$/lb.

4

DESIGN OF SLENDER BEAMS FOR MINIMUM WEIGHT

A slender member subjected primarily to transverse loads is termed a beam. Generally at a given cross section such a member must carry both bending moment and direct shear. In this chapter we will limit our study to the design of an optimum (minimum volume) cross section required to carry a given bending moment. A cross section designed on this basis can be modified to account for the presence of direct shear. For example, the thickness of web elements can be increased based on interaction data for combined bending and shearing stresses. This "after the fact" approach obviously prevents a redistribution of material from flange elements to web elements which would occur in a simultaneous consideration of bending moment and direct shear.* Since typically the modification of web thickness is small, the redistribution of material is too small to be concerned with. In other words, a beam optimized to carry pure bending and then modified for the shear-carrying requirement represents a compromise between the true optimum and analytical feasibility, resulting in a "near-optimum" design.

Section 1

ASSESSMENT OF THE ENVIRONMENT

So that we can analyze bending using engineering beam theory, we will limit our consideration to cross sections with an axis of symmetry normal to the bending moment vector as shown in Fig. 4-1 on p. 64. Although the environment here implies no specific failure mode (as was the case in the column problem where Euler buckling was implied by environment alone), there is some insight to be gained by considering the state of applied stress as a func-

*An increased web thickness, in addition to carrying the shear load, would increase the moment of inertia, thus allowing for a possible reduction in flange material.

tion of given environment and cross section properties. For linear materials the maximum bending stress can be expressed by the well-known flexure equation

$$\sigma_{A\max} = \frac{M_b C_{\max}}{I_{xx}} \qquad (4\text{-}1)$$

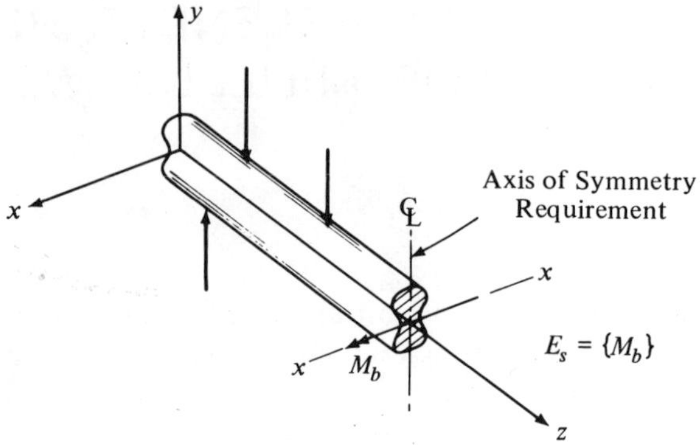

Fig. 4-1. Environment and Cross Section Limitation

It can be seen, therefore, that for a given bending moment, the maximum applied normal stress will be smallest when I_{xx}/C^* is a maximum. In the direction of minimum volume this can be achieved most efficiently by locating the bulk of material symmetrically away from the bending axis and providing a web shear tie such as we have in the H-section. In order to introduce the optimization approach for beams, however, we will first consider a two-variable system in the form of a circular tube.

Section 2

THE CIRCULAR TUBE BEAM

Establishing the merit function as a minimum in weight per unit length of beam, we have

$$\rho A = \rho \pi D t \longrightarrow \min \qquad (4\text{-}2)$$

The applied maximum normal stress can be evaluated as a function of environment and system variables as follows:
For a circular tube

$$I = \frac{\pi D^3 t}{8}$$

*Referred to as the section modulus.

Therefore from Eq. (4-1)

$$\sigma_A = \frac{M_b(D/2)}{\frac{\pi D^3 t}{8}} = \frac{4M_b}{\pi D^2 t} \qquad (4\text{-}3)$$

The inherent failure modes for a circular tube in bending are excessive stress* and local buckling of the compression surface. Accordingly we have for our failure constraints

$$\sigma_A \leq \sigma_{PL} \qquad (4\text{-}4)$$
$$\sigma_A \leq \sigma_L \qquad (4\text{-}5)$$

The local buckling stress, σ_L, as a function of material and geometry can be taken as that for a circular tube under uniform axial compression stress since in bending a significant portion of the circumference is subjected to a relatively uniform compression field. Accordingly, Eq. (2-7), deleting the inelastic correction $\eta_T^{1/2}$, applies as follows

$$\sigma_A = \psi_L K_c E\left(\frac{t}{D}\right) \qquad (4\text{-}6)$$
$$\psi_L \leq 1$$

as the local buckling constraint. Introducing a slack-variable into Eq. (4-4) yields, for the excessive stress failure constraint,

$$\sigma_A = \psi_{PL} \sigma_{PL} \qquad (4\text{-}7)$$
$$\psi_{PL} \leq 1$$

Inserting the expression for applied stress, Eq. (4-3), into the above constraints results in

$$\frac{4M_b}{\pi D^2 t} = \psi_L K_c E\left(\frac{t}{D}\right) \qquad (4\text{-}8)$$

$$\frac{4M_b}{\pi D^2 t} = \psi_{PL} \sigma_{PL} \qquad (4\text{-}9)$$

Our problem now is to find a stationary minimum for Eq. (4-2) subject to satisfaction of Eqs. (4-8) and (4-9). Recall that in the column problem the excessive stress constraint was not written as a conditional equation containing a slack variable. There, by employing an equivalent applied stress merit (unique to axially loaded members), the excessive stress constraint simply established an upper bound on a maximized applied stress merit, subject to the local and Euler constraints. For the present case no conversion exists for which the applied stress becomes an equivalent merit function. As a

---

*Employing the flexure equation limits our stress values to the linear portion of the stress–strain diagram. Therefore, our upper bound stress will be taken as the proportional limit stress except for sharply yielding materials such as mild steel, in which case the proportional limit and yield stress become coincident.

consequence, we must deal directly with the cross-sectional area in seeking an optimum. This presents no particular problem and is discussed here only to parallel the following evaluation with the approach of Chap. 2.

The general approach requires algebraic combination of merit and constraints in the direction of reducing the number of proportion variables in the merit function. Once the slack-variables have been introduced into the merit function, they, in addition to any remaining proportion variables, can be optimized with regard to their effects on merit.

Toward this end, simultaneous solution of Eqs. (4-8) and (4-9) results in both t and D expressed as functions of environment, material variables, and slack-variables as follows:

$$t = \left(\frac{4\psi_{PL}\sigma_{PL}M_b}{\pi K_c^2 \psi_L^2 E^2}\right)^{1/3} \tag{4-10}$$

$$D = \left(\frac{4\psi_L K_c E M_b}{\pi \psi_{PL}^2 \sigma_{PL}^2}\right)^{1/3} \tag{4-11}$$

Now, inserting Eqs. (4-10) and (4-11) into the merit function, Eq. (4-2), yields

$$\rho A = \frac{4^{2/3}\pi^{1/3}\rho M_b^{2/3}}{\psi_L^{1/3}\psi_{PL}^{1/3}K_c^{1/3}E^{1/3}\sigma_{PL}^{1/3}} \longrightarrow \min \tag{4-12}$$

The inverse product dependence of ρA on ψ_L and ψ_{PL} demonstrates that for minimum weight

$$\psi_{L_{\text{opt}}} = \psi_{PL_{\text{opt}}} = 1 \tag{4-13}$$

and thus the optimum tube will be a SMD. Evaluating Eq. (4-12) at this condition with $K_c = 0.40$ (see the discussion below Eq. (2-7) in Chap. 2) results in

$$(\rho A)_{\text{opt}} = 5.00 \left(\frac{\rho}{E^{1/3}\sigma_{PL}^{1/3}}\right) M_b^{2/3} \tag{4-14}$$

and identifies

$$\frac{E^{1/3}\sigma_{PL}^{1/3}}{\rho} \longrightarrow \max \tag{4-15}$$

as the material metric.

The load index in a tapered beam design can be considered the local value of M, or, for a uniform beam, the maximum value of M. For the *uniform* beam of length L, an index representation is often desired where the load index has the units of stress.* This is accomplished by first writing Eq. (4-14) as a total weight expression

---

*It is shown in Chap. 5 that a load index in units of stress can facilitate the comparison between alternative system forms for the general environment.

$$W_{opt} = 5.00\left(\frac{\rho}{E^{1/3}\sigma_{PL}^{1/3}}\right)M_b^{2/3}L \qquad (4\text{-}16)$$

We now see that division of both sides of Eq. (4-16) by L^3 results in a load index of (M_b/L^3) in units of stress, thus:

$$\left(\frac{W}{L^3}\right)_{opt} = 5.00\left(\frac{\rho}{E^{1/3}\sigma_{PL}^{1/3}}\right)\left(\frac{M_b}{L^3}\right)^{2/3} \qquad (4\text{-}17)$$

and identifies W/L^3 as a corresponding weight index. Note that Eq. (4-17) applies only for a beam of uniform cross section along its length and hence M_b must be taken as the maximum moment.

Section 3

THE H-SECTION BEAM

Employing the same form and variables of Fig. 2-7, and establishing the merit function as a minimum in weight per unit length, we have

$$\rho A = \rho t h(1 + 2k_1 k_2) \longrightarrow \min \qquad (4\text{-}18)$$

The applied maximum normal stress can be evaluated as a function of environment and system variables as follows:
About the bending axis (x-x axis)

$$I_{xx} = \frac{th^3}{12}(1 + 6k_1 k_2) \qquad (4\text{-}19)$$

Therefore from Eq. (4-1)

$$\sigma_A = \frac{6M_b}{th^2(1 + 6k_1 k_2)} \qquad (4\text{-}20)$$

The inherent failure modes for the H-section in bending are excessive stress in the flange and local buckling of the flange and web. Accordingly our failure constraints become

$$\sigma_A \leq \sigma_{PL} \qquad (4\text{-}21)$$
$$\sigma_A \leq \sigma_{Lf} \qquad (4\text{-}22)$$
$$\sigma_A \leq \sigma_{Lw} \qquad (4\text{-}23)$$

The critical buckling stress for the flange is due to plate buckling under uniform stress with one longitudina ledge free and the other simply supported. Therefore*

$$\sigma_{Lf} = 0.385E\left(\frac{2k_2 t}{k_1 h}\right)^2 \qquad (4\text{-}24)$$

*See Eq. (2-26).

For the web with both longitudinal edges simply supported and sustaining a linearly varying tension-compression field, the buckling equation (based on the maximum stress in the web) takes the form[11]

$$\sigma_{L_w} = 21.7E\left(\frac{t}{h}\right)^2 \tag{4-25}$$

As in the case of the *H*-section column we will allow for the possibility of a conservative flange with respect to local buckling (see the discussion in Chap. 2 below Eq. (2-29)) by stipulating

$$\sigma_{L_w} \leq \sigma_{L_f} \tag{4-26}$$

Accordingly, combining Eqs. (4-24), (4-25), and (4-26) yields the following constraining relation for k_1 and k_2

$$k_2 \geq 3.75 k_1 \tag{4-27}$$

As a result, we can select as our critical local failure mode Eq. (4-25), and, introducing a slack-variable, we have

$$\sigma_A = \psi_L 21.7E\left(\frac{t}{h}\right)^2 \tag{4-28}$$

$$\psi_L \leq 1$$

To complete the failure constraints, we insist that

$$\sigma_A = \psi_{PL}\sigma_{PL} \tag{4-29}$$

$$\psi_{PL} \leq 1$$

Now, inserting the equation for applied stress, Eq. (4-20), into the above constraints yields

$$\frac{6M_b}{th^2(1 + 6k_1k_2)} = \psi_L 21.7E\left(\frac{t}{h}\right)^2 \tag{4-30}$$

$$\frac{6M_b}{th^2(1 + 6k_1k_2)} = \psi_{PL}\sigma_{PL} \tag{4-31}$$

Simultaneous solution of the above equations results in the following relationships for t and h

$$t = \left[\frac{6M_b}{21.7\psi_L E(1 + 6k_1k_2)}\right]^{1/3} \tag{4-32}$$

$$h = \left[\frac{6^{2/3}M_b^{2/3}\psi_L^{1/3}(21.7E)^{1/3}}{\psi_{PL}\sigma_{PL}(1 + 6k_1k_2)^{2/3}}\right]^{1/2} \tag{4-33}$$

Substituting these into Eq. (4-18) yields the merit function in terms of environment, material variables, and the slack-variables as follows (also in terms of the two remaining proportion variables, k_1 and k_2)

Section 3 — The H-Section Beam

$$pA = \frac{(1 + 2k_1k_2)}{(1 + 6k_1k_2)^{2/3}} \frac{6^{2/3}}{(21.7)^{1/6}} \frac{p}{E^{1/6}\sigma_{PL}^{1/2}} \frac{M_b^{2/3}}{\psi_L^{1/6}\psi_{PL}^{1/2}} \longrightarrow \min \quad (4\text{-}34)$$

The dependence of area on an inverse product form of ψ_L and ψ_{PL} demonstrates that $\psi_{L_{\text{opt}}} = \psi_{PL_{\text{opt}}} = 1$, and hence local buckling of the web and excessive stress occur simultaneously at the applied moment. Whether the flange buckles with the web or is conservative with respect to this failure mode can only be determined once optimum values of k_1 and k_2, subject to the constraint of Eq. (4-27), have been evaluated. Recall that the equality of Eq. (4-27) implies simultaneous occurrence of web and flange buckling, whereas the inequality implies a conservative flange.

Since Eq. (4-34) depends on k_1 and k_2 as the product k_1k_2, it follows that area will be minimum when

$$f(\theta) = \frac{6^{2/3}(1 + 2\theta)}{(21.7)^{1/6}(1 + 6\theta)^{2/3}} \longrightarrow \min$$

where

$$\theta = k_1k_2$$

Forming $df/d\theta = 0$ yields

$$\frac{(1 + 6\theta) - 2(1 + 2\theta)}{(1 + 6\theta)^{5/3}} = 0 \quad (4\text{-}35)$$

from which $\theta_{\text{opt}} = (k_1k_2)_{\text{opt}} = \tfrac{1}{2}$.

Hence Eq. (4-34) will be minimum when the product k_1k_2 is $\tfrac{1}{2}$. This results in a family of optimum pairs of k_1 and k_2 all yielding the same minimum weight, subject to the restriction on permissible pairs of k_1 and k_2 as given in the constraint of Eq. (4-27). Accordingly the optimum family $(k_1, k_2)_{\text{opt}}$ can be visualized as shown in Fig. 4-2.

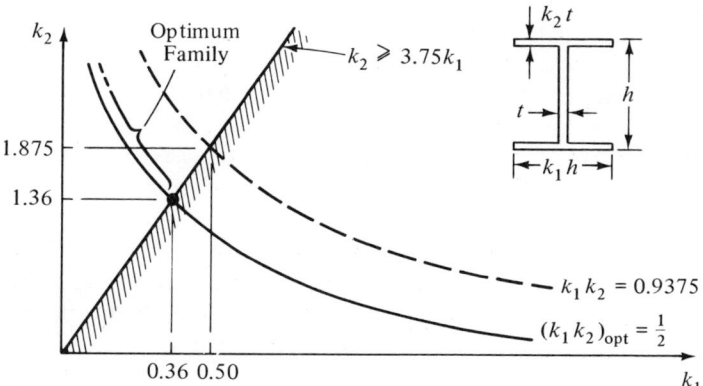

Fig. 4-2. Optimum Pairs of k_1 and k_2

Note that all points above the line $k_2 = 3.75\ k_1$ and on the hyperbola $k_1 k_2 = \frac{1}{2}$ are optimum selections, *but are not simultaneous mode designs.**
Note also that this limits the value of k_1 to be less than 0.36. To visualize this, consider the proportions of the SMD where $k_1 = 0.36$ ($k_2 = 1.36$) as shown in Fig. 4-3.

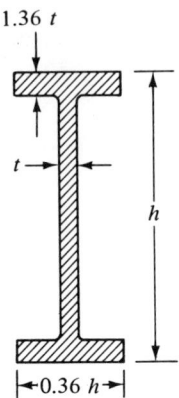

Fig. 4-3. Proportions of the SMD Cross Section

Substituting the SMD optimum values of k_1 and k_2 into Eq. (4-34) yields

$$(\rho A)_{\text{opt(SMD)}} = 1.565 \left(\frac{\rho}{E^{1/6} \sigma_{PL}^{1/2}} \right) M_b^{2/3} \quad (4\text{-}36)$$

It would be desirable to have a flange width which was at least one-half the web depth ($k_1 = \frac{1}{2}$) in order to prevent "lateral buckling," as discussed by Timoshenko.[11] This can be achieved by selecting $k_1 k_2 = 0.9375$ as shown in Fig. 4-2 by the dashed curve. For this "off-optimum" selection of the product $k_1 k_2$, k_1 must be less than or equal to 0.500, and thus we have achieved the desired modification. The question remains as to how much the area has been increased as a consequence.

To answer the question above we can perform a "sensitivity" analysis of the form ρA as a function of the product $k_1 k_2$. This is shown in Fig. 4-4 and represents the plot of $f(\theta)$ as a function of $k_1 k_2$.

Note that selecting the off-optimum value of $k_1 k_2 = 0.9375$ results in about a 2 percent increase in cross-sectional area. It would obviously be desirable to accept this weight penalty in favor of the "better-proportioned" cross section which would result as shown in Fig. 4-5.

*The nonsimultaneous mode designs correspond to cross sections for which the thickness-to-width ratio of the flange is increased while keeping the flange area constant.

Section 3   The H-Section Beam   71

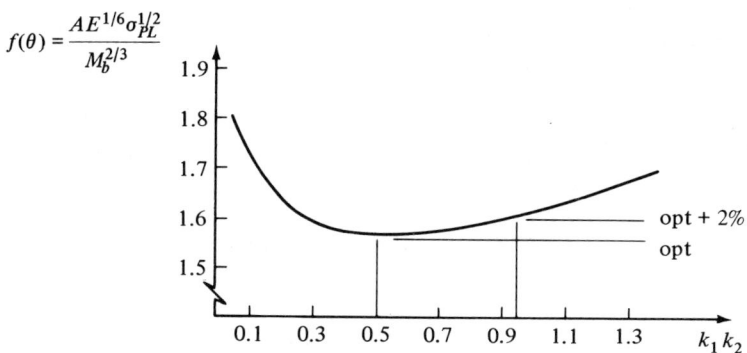

Fig. 4-4. Sensitivity Analysis w.r.t. $(k_1 k_2)$

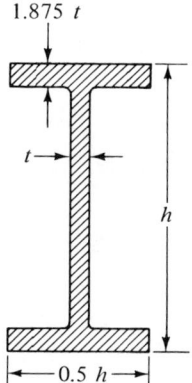

Fig. 4-5. "Off-optimum" Proportions for $k_1 k_2 = 0.9375$ and Weight Penalty = 2 percent

Substituting $k_1 = .500$ and $k_2 = 1.875$ into Eq. (4-34) yields

$$(\rho A)_{(opt+2\%)} = 1.606 \left(\frac{\rho}{E^{1/6} \sigma_{PL}^{1/2}} \right) M_b^{2/3} \qquad (4\text{-}37)$$

$$\text{for} \begin{cases} k_1 = 0.500 \\ k_2 = 1.875 \end{cases}$$

Equations (4-32) and (4-33) are the necessary design equations for the determination of t and h for a given bending moment and material selection. From Eq. (4-37) we identify the material metric as

$$\frac{E^{1/6} \sigma_{PL}^{1/2}}{\rho} \longrightarrow \max \qquad (4\text{-}38)$$

Section 4
GEOMETRIC CONSTRAINTS

In a given problem, Eqs. (4-10) and (4-11) can be employed to evaluate t_{opt} and D_{opt} at the optimum values of ψ_L and ψ_{PL} for a circular tube. Suppose, however, that it is required to fix the diameter at a value which exceeds D_{opt}, such as might be the case in a fuselage shell.

Consider first the dependence of D on ψ_L and ψ_{PL} in Eq. (4-11), reproduced below for convenience

$$D \propto \frac{\psi_L^{1/3}}{\psi_{PL}^{2/3}} \qquad (4\text{-}39)$$

Under the constraints it can be seen that an increased diameter can only be achieved by selecting a value for ψ_{PL} which is less than unity, resulting in a conservative margin with respect to failure by excessive stress. The value of ψ_{PL} necessary for a constrained value for D ($D_{co} > D_{opt}$) must be

$$(\psi_{PL})_{co}^{2/3} = \frac{D_{opt}}{D_{co}} \qquad (4\text{-}40)$$

where the subscript co denotes a constrained optimum. The corresponding value of t_{co} can then be determined from Eq. (4-10) and results in

$$t_{co} = \left(\frac{4 \left(\frac{D_{opt}}{D_{co}} \right)^{3/2} \sigma_{PL} M_b}{\pi K_c^2 E^2} \right)^{1/3} \qquad (4\text{-}41)$$

The weight penalty can be determined for the above value of ψ_{PL} from Eq. (4-12) and results in

$$\frac{(\rho A)_{co}}{(\rho A)_{opt}} = \frac{1}{\psi_{PL}^{1/3}} = \left(\frac{D_{co}}{D_{opt}} \right)^{1/2} \qquad (4\text{-}42)$$

For example, if D_{co} is fixed at $2D_{opt}$, Eq. (4-42) yields a weight ratio of 1.414, a 40 percent weight penalty. Large ratios of D_{co}/D_{opt} are common in the design of large (specified) diameter fuselage shell structures in aircraft applications. In such cases a simple monocoque shell is seldom a strong candidate. The remedy for these large weight penalties can be inferred from the above analysis. Note that the effect of a constraint on D ($D_{co} > D_{opt}$) is to yield a conservative margin against failure by excessive stress. What this means is that the large specified diameter shell buckles before it reaches the maximum stress level. This suggests the idea that associated weight penalties could be combated by stiffening the shell both longitudinally and circumferentially so as to increase the stress for which the shell buckles. We shall explore this fundamental approach to the design of large specified diameter tubes under bending in Chap. 7.

Section 5

TAPERED BEAMS

In this section we consider the weight advantage of tapering the proportion variables such that the optimum cross section changes continuously with variations in the bending moment.

For the general beam, the minimum value of cross-sectional area at any point is proportioned to $M^{2/3}$. If we arbitrate a uniform beam over the entire length, the cross-sectional area must be based on the maximum bending moment and hence the total beam weight would be proportional to $M_{max}^{2/3}L$. For a tapered beam, the weight of a differential length of beam, dx, would be proportional to $M^{2/3}(x)\,dx$, where here $M(x)$ is the current value of moment. Over the entire length of a tapered beam the total weight would therefore be proportional to $\int_0^L M^{2/3}(x)\,dx$. Hence the ratio of the uniform beam's weight to that of the tapered design can be expressed as

$$R_{u/T} = \frac{M_{max}^{2/3} L}{\int_0^L M^{2/3}(x)\,dx} \qquad (4\text{-}43)$$

For example, consider a single load, P, cantilevered a distance L from a wall. If we establish the origin at the point of applied load, the moment at any point is simply $M(x) = Px$, and the maximum moment is $M_{max} = PL$. Employing Eq. (4-43) with these expressions yields

$$R_{u/T} = \frac{(PL)^{2/3} L}{\int_0^L (Px)^{2/3}\,dx} = 1.67 \qquad (4\text{-}44)$$

and thus the uniform beam is seen to be 67 percent heavier then a tapered design for this case. For the same cantilevered environment, but with a uniform load distribution, q, over the entire length, the moment at any point can be expressed as $M(x) = qx^2/2$ with $M_{max} = qL^2/2$. Therefore Eq. (4-43) yields

$$R_{u/T} = \frac{(qL^2/2)^{2/3} L}{\int_0^L (qx^2/2)^{2/3}\,dx} = 2.34 \qquad (4\text{-}45)$$

Here the weight difference is 134 percent. Such significant weight savings are peculiar to the cantilever environment since, under this type of external restraint, the moment function falls relatively fast from its maximum value at the built-in wall to zero at the free end.

To illustrate that such weight savings are less dramatic in a noncantilever environment, consider a beam simply supported at both ends and supporting a uniform load distribution, q. The moment at any point is $M(x) = (qLx - qx^2)/2$ and $M_{max} = qL^2/8$. Therefore

$$R_{u/T} = \frac{(qL^2/8)^{2/3}L}{\int_0^L (qLx/2 - qx^2/2)^{2/3}\,dx} \tag{4-46}$$

This integral form is nonelementary. An approximate evaluation employing Simpson's rule with four equal intervals $(L/4)$ results in $R_{u/t} = 1.32$ for this case.

PROBLEMS

4-1. Compute the weight ratio of a circular tube beam to an H-section beam for (a) aluminum 7075-T6 (b) AISI 1025 steel (Assume $\sigma_{PL} = \sigma_y$).
Answer (a) 1.38, (b) 1.02.

4-2. Optimize the rectangular tube cross section shown.

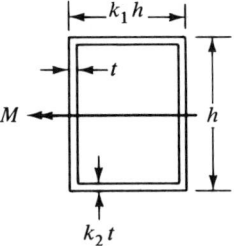

Assume simple edge supports at the web-flange intersection and allow for a possibly conservative flange, vis., $\sigma_{L_w} \leq \sigma_{L_f}$.
Show that a family solution results of the form

$$(\rho A)_{opt} = \frac{1.198\rho[1 + (k_1 k_2)_{opt}]}{E^{1/6}\sigma_{PL}^{1/2}[\frac{1}{3} + (k_1 k_2)_{opt}]^{2/3}} M_b^{2/3} = \rho \frac{1.97 M_b^{2/3}}{E^{1/6}\sigma_{PL}^{1/2}}$$

where

$$(k_1 k_2)_{opt} = 1, \quad k_{1\,opt} \leq 0.408 k_{2\,opt}$$

4-3. Show that, independent of material and applied moment, an optimized rectangular tube is 26 percent heavier than an optimized H-section.

4-4. Evaluate the optimum proportion variables for an H-section beam of AISI 1025 steel under a bending moment of 100,000 in-lb, (a) at the SMD ($k_1 = 0.36$, $k_2 = 1.36$), (b) at $k_1 = 0.25$ and $k_2 = 2.00$. Show that both yield the same weight, and that in both cases webs are identical and flanges have equal areas with different thickness-to-width ratios.

4-5. Show that, independent of material and applied moment, an optimized solid circular cross section is 12 percent heavier than an optimized solid square cross section. How does the solid square compare with the H-section?

4-6. Show that the optimum weights for beams under the given environments can be expressed in the form shown. Make the evaluation for uniform H-section beams.

$$\frac{W}{L^3} = \frac{1.606\rho}{E^{1/6}\sigma_{PL}^{1/2}}\left(\frac{M_0}{L^3}\right)^{2/3} \quad \frac{W}{L^3} = \frac{1.606\rho}{(4)^{2/3}E^{1/6}\sigma_{PL}^{1/2}}\left(\frac{P}{L^2}\right)^{2/3} \quad \frac{W}{L^3} = \frac{1.606\rho}{(8)^{2/3}E^{1/6}\sigma_{PL}^{1/2}}\left(\frac{q}{L}\right)^{2/3}$$

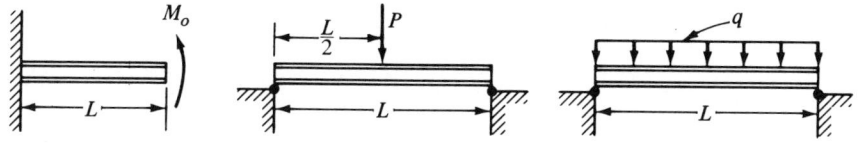

4-7. As opposed to carrying a moment M with one optimized circular tube, it is proposed to carry $M/2$ each with two tubes side by side as shown.

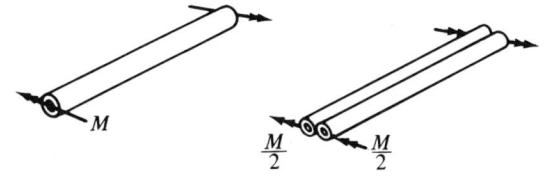

Show that this must always result in a weight penalty (26.5 percent).

4-8. Use graphical design space to evaluate D_{opt} and t_{opt} for an aluminum 7075-T6 (assume $\sigma_{PL} = \sigma_y$) circular tube beam under a bending moment of 100,000 in-lb. Impose an arbitrary constraint which doubles D_{opt} and determine the corresponding constrained optimum value of t. Compare this to the analytical solution.

4-9. Determine the weight ratio, $R_{u/T}$, for a simply supported beam with a concentrated load, P, at the center span.

5

DESIGN OF STRUCTURAL SYSTEMS

The slack variable approach to the inequality constrained optimization problem has been applied in previous chapters to single-element forms involving at most four unknown proportion variables. If we were to specify a form containing a number of such elements it would appear that the proportionate increase in the number of unknown variables and failure constraints would overwhelm a parametric attempt to optimize employing a similar approach. Obviously each element of a multiple-element structure cannot be optimized independently. The orientation variables* define a load state for each element which varies as the orientation variables are changed. To say that each element could be optimized independently would in effect be saying that the orientation variables could be fixed arbitrarily and would fail to take advantage of the possibility that total merit might depend critically on orientation.

What is needed is a systems approach which utilizes the parametric optimization results for single elements but at the same time allows for the orientation variables to be optimized.

Section 1
SYSTEMS AS COMBINATIONS OF OPTIMIZED ELEMENTS

For three fundamental modes of load transmission, axial tension, axial compression, and bending moment, relationships have been developed which define optimum proportions for various single-element forms as a function

---

*Recall that in addition to the respective proportion and material variables (S_{pi}, S_{mi}) of the system elements, a structural system will require an array of orientation variables (S_o) which define the spacial coordination between elements in terms of angles, lengths, and ratios (See Sec. 6, Chap. 1).

of material variables and local environment.* These relationships can now be viewed as mathematical "building blocks" in the design of structural forms for which the system environment to be sustained does not *directly* manifest itself in the load-geometry environment of either axial force or bending moment transmission. (Consider as an example the environment of Fig. 1-3.) A configuration of the basic load transmission elements of cables, columns, and beams can often constitute a structure which supports a given load environment more efficiently than any of the elements acting alone. With this in mind we now explore the next logical extension in the search for optimum form: the multiple-element structural system.

In what follows, a general systems design approach will be developed and applied. We will see that the design of structural systems composed of previously optimized elements can be reduced to an evaluation of a stationary merit function, where the only independent variables to be optimized are the orientation variables and redundant loads (in the sense of statically indeterminate) inherent in a specified system form. The ability to consider the optimum values of all proportion variables as having dependence on orientation and redundancy will result in the potential to optimize structural systems composed of large numbers of system variables while dealing with a mathematical optimization problem involving only a few independent variables. In addition, employing the parametric optimization results developed for component elements satisfies the inequality failure constraints, in which case the systems optimization problem is further reduced to an unconstrained evaluation with the optimum condition obtained through ordinary calculus.

We will also see that system weight results can be reduced to index form with respect to dependence on environment. In this way comparisons between alternative system forms will be effected in the same manner as those previously made between alternative single-element forms.

Section 2
SYSTEM TYPES

We can classify structural systems as either statically determinate or statically indeterminate. For compactness of terminology we will refer to these respectively as *determinate* and *redundant*.

*In a following section we shall make an important distinction between local environment and system environment. Suffice it to say here that local environments are the loads and transmission lengths associated with particular elements of a structural system, whereas the system environment is the array of fixed environmental parameters (E_s) manifest before form is specified.

A determinate system is a spacial configuration of structural elements, mutually connected and externally restrained such that by the equations of equilibrium alone, the loads carried by the respective elements can be determined in terms of orientation variables and system environment.

A redundant system contains extraneous (redundant) elements or external restraints such that, by employing the equations of equilibrium alone, the loads carried by the respective elements must be expressed in terms of redundant loads in addition to the orientation variables and system environment. We say that a structure is N-redundant if it contains N such redundant loads.

At first glance it might seem that structural redundancy would present considerable mathematical difficulty in a design problem. Recall that in the analysis problem of a structure with given proportions and orientation, either compatibility of deformation or an energy principle must be employed in order to supplement the equations of equilibrium. With this supplementation, the redundant loads are evaluated in terms of the respective stiffnesses of the elements, viz., the proportion and material variables. Since in the design problem the proportion and material variables are the output of the evaluation, not the input, insurmountable mathematics would seem present in all but the most trivial redundant design problems. Actually the analytical design of redundant structures will be shown to present very little difficulty over that for determinate structures. As noted in the previous section, the general approach will allow for redundant loads to be optimized within the merit function in the same manner as orientation variables. Once orientation variables and redundant loads have been determined, the proportion variables will follow as dependent quantities. Compatibility can also be shown to follow and is not, therefore, employed in the design evaluation.* Applications in this chapter will be with determinate structures, leaving applications to redundant structures for Chap. 6.

The following classifications of system types will also be useful in subsequent application: (a) *similar elements*, (b) *dissimilar elements*, (c) *prestressed* (the structure sustains a self-equilibrating set of internal loads in the unloaded state), (d) *composite* (the structure utilizes a heterogeneous cross section of dissimilar materials).

A prestressed structure is inherently redundant and hence applications are considered in Chap. 6. Applications to composites will be taken up in Chap. 7.

---

*See Sec. 4, Chap. 6.

Section 3
THE LOCAL ENVIRONMENT

As a first step in modifying the optimization phase of design discussed in Chap. 1 so that it more directly applies to systems, we make the following definition:

> The local environment for the ith element is the array of parameters which define the loads and transmission lengths for the particular element.

We can functionally express these parameters in terms of the fixed system environment (E_s), the orientation variables (S_o), and redundant loads (R_N), viz.

$$E_{si} = E_{si}(E_s, S_o, R_N) \qquad (5\text{-}1)$$

The functional relationships implied by Eq. (5-1) are obtained by a statical and geometrical analysis of the specified system form. To illustrate this, consider first the statically determinate system of Fig. 5-1. We know from past optimization studies of axially loaded elements that knowledge of axial force and axial length magnitudes represents sufficient environmental information for determining the optimum (minimum-weight) proportions. These parameters, therefore, are by definition the local environment. Note in the figure that from the considerations of statics and geometry, the two local

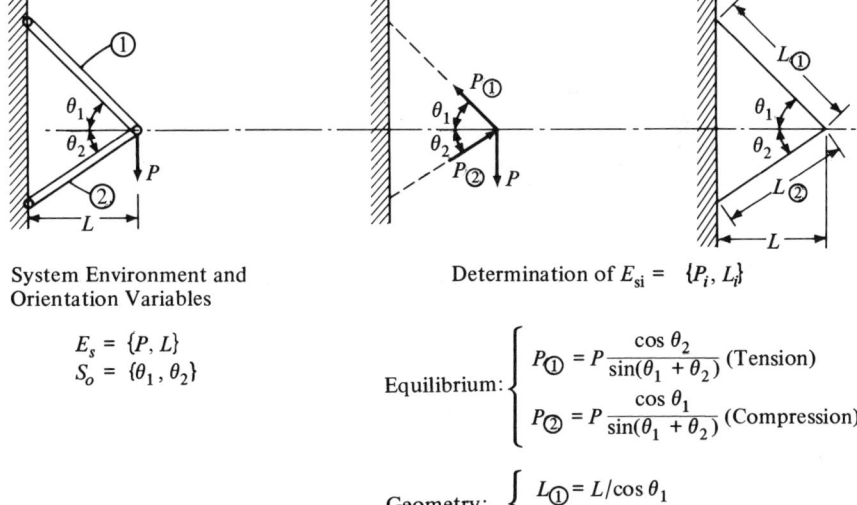

Fig. 5-1. Local Environments for a Statically Determinate Truss

environments have been expressed in terms of the fixed system environment and the orientation variables.

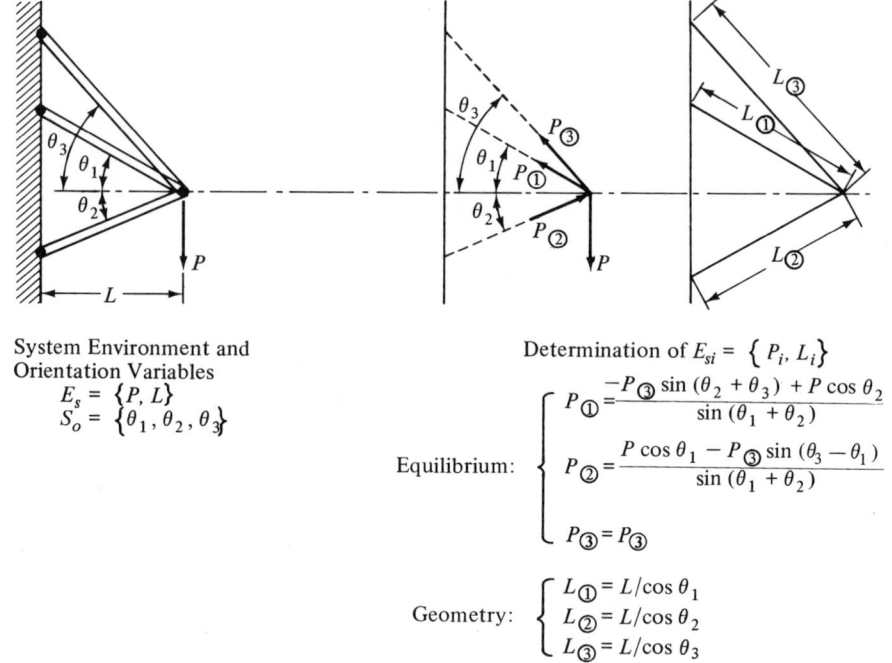

Fig. 5-2. Local Environments for a Statistically Indeterminate Truss

We next consider a one-redundant three-bar truss as shown in Fig. 5-2. The three local environments must now be expressed in terms of the redundant load ($P_③$) in addition to the fixed system environment and orientation variables.

Note that the local environment evaluations in Fig. 5-1 and 5-2 are consistent with the form of Eq. (5-1).

Section 4

ANALYTICAL APPROACH TO SYSTEMS DESIGN

Consider a general multiple-element system. Suppose that for the ith element an optimized merit function has been evaluated parametrically in terms of its local environment and material variables as shown

$$M_{i_{opt}} = M_i(E_{si}, S_{mi}) \qquad (5\text{-}2)$$

If individual merit functions exist in the form of Eq. (5-2), the merit for the system can be expressed as

$$M_s = \sum_i M_i(E_{si}, S_{mi}) \tag{5-3}$$

provided the individual merit functions are additive.*

The optimum subscripts in Eq. (5-2) imply an optimized condition for given (fixed) values of E_{si} and S_{mi}. Since in systems design the E_{si}'s will functionally depend on S_o and R_N, we do not show an optimum subscript for M_s in Eq. (5-3), pending determination of $\{S_o\}_{opt}$ and $\{R_N\}_{opt}$. Towards this end, employing the generalization of Eq. (5-1) combined with Eq. (5-3) yields

$$M_s = \sum_i M_i(E_s, S_o, R_N, S_{mi}) \tag{5-4}$$

The functional form of Eq. (5-4) demonstrates that, in a system design for which each element has been optimized in terms of its local environment, the only independent variables are the orientation variables and redundant loads.† By this approach, the optimum determination of all proportion variables is seen to be a dependent evaluation. The requirement for an optimum system can now be stated simply as

$$\frac{\partial}{\partial S_o}\left\{\sum_i M_i(E_s, S_o, R_N, S_{mi})\right\} = 0 \tag{5-5}$$

$$\frac{\partial}{\partial R_N}\left\{\sum_i M_i(E_s, S_o, R_N, S_{mi})\right\} = 0 \tag{5-6}$$

for each S_o and R_N.

The search for $\{S_o, R_N\}_{opt}$ is unconstrained since all practically realizable values of S_o and R_N are acceptable designs.

Section 5

DETERMINATE SIMILAR-ELEMENT SYSTEMS

As a first example, consider the system environment of Fig. 5-3(a). A load P is to be carried at a vertical distance L from the vertex of a roof with vertex

---

*By additive we mean simply that the merit function for the total system is equal to the sum of the merit functions for individual elements. If the elements merit functions are either all cost or all weight, the additive property obviously holds. For the "equivalent" merit function of $(\sigma_A/\rho) \rightarrow$ max for minimum weight of axially loaded members, the additive property does not hold. In this case the merit function must be converted to a weight expression before Eq. (5-3) applies.

† In some cases a general buckling mode failure exists for the total structure which is not accounted for in the component element optimization results. Here the search for optimum S_o and R_N would be subject to this single constraint. An example would be an elongated truss under longtitudinal end compression.

angle 2ϕ. We accordingly recognize the fixed system environment as $E_s = \{P, L, \phi\}$.

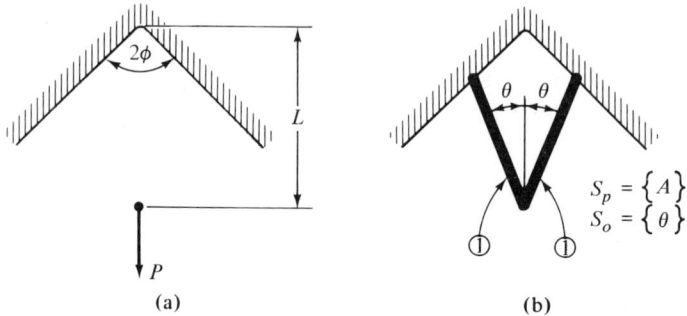

Fig. 5-3. Two-cable System with Variable Orientation

We shall specify a determinate similar-element system in the form of two symmetrical axial cables whose orientation is defined by the angle between the cable and the vertical as shown in Fig. 5-3(b). Our problem will be to determine the cable cross-sectional area ($S_p = A$) and orientation ($S_o = \theta$) for an established merit function of minimum total weight.

We know from previous considerations (see Example 1-1, Eq. (1-16)) that a member in axial tension will be of minimum weight for parametric values of local environment and material variables as follows

$$W_{①_{opt}} = \frac{\rho}{\sigma_y} P_① L_① \tag{5-7}$$

Since the system is symmetrical, the total system weight can be expressed as

$$W_s = 2\frac{\rho}{\sigma_y} P_① L_① \tag{5-8}$$

Note that Eq. (5-7) carries an optimum subscript since in its derivation it was assumed that $P_①$ and $L_①$ represent fixed environment. In the present system combination of this element we recognize that $P_①$ and $L_①$ are local environments dependent on orientation and hence no optimum subscript is shown in Eq. (5-8) pending a determination of θ_{opt}.

The local environment is evaluated in terms of P, L, ϕ, and θ by a consideration of statics and geometry as shown in Fig. 5-4. From equilibrium of concurrent forces it is found that

$$P_① = \frac{P}{2 \cos \theta} \tag{5-9}$$

Also, the Law of Sines yields

$$L_① = \frac{L \sin \phi}{\sin (\phi + \theta)} \tag{5-10}$$

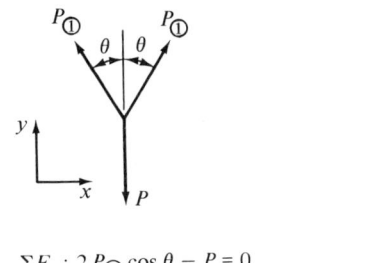
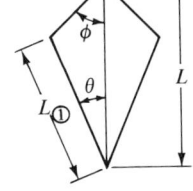

$\Sigma F_y : 2P_{①} \cos \theta - P = 0$

$$\frac{L}{\sin[180 - (\phi + \theta)]} = \frac{L_{①}}{\sin \phi}$$

$$\sin[180 - (\phi + \theta)] \equiv \sin(\phi + \theta)$$

Fig. 5-4. Evaluation of Local Environment for Two-cable System

By combining Eqs. (5-8), (5-9), and (5-10) we can express total system weight in terms of the orientation variable, θ, as follows

$$W_s = \frac{\rho}{\sigma_y} PL \left[\frac{\sin \phi}{\cos \theta \sin(\phi + \theta)} \right] \qquad (5\text{-}11)$$

We are now in a position to determine θ_{opt} such that the total system weight is a minimum.* Furthermore, since $W_s \to$ min when

$$f_e(\theta) = \left[\frac{\sin \phi}{\cos \theta \sin(\phi + \theta)} \right] \to \text{min} \qquad (5\text{-}12)$$

we see that θ_{opt} is independent of P, L, ρ, and σ_y. θ is therefore seen to be a secondary system variable with respect to these quantities. There will, however, be dependence on the roof angle ϕ. To establish this dependence we form

$$\frac{df_e}{d\theta} = 0$$

which yields

$$\cot(\theta + \phi) = \tan \theta$$

or

$$\theta + \phi = 90 - \theta$$

and therefore

$$\theta_{opt} = \frac{90 - \phi}{2} \qquad (5\text{-}13)$$

*Dividing both sides of Eq. (5-11) by L^3 yields a convenient index formulation of the weight result, where the environment has been lumped into a load index of P/L^2, viz.,

$$\frac{W_s}{L^3} = \frac{\rho}{\sigma_y} \left[\frac{\sin \phi}{\cos \theta \sin(\phi + \theta)} \right] \left(\frac{P}{L^2} \right)$$

It is in terms of this "indicized" evaluation that subsequent system form comparisons can most conveniently be made with respect to the general environment of Fig. 5-3(a).

Now as a dependent evaluation from Eqs. (5-9) and (5-13) we determine the optimum proportion variable as

$$A_{opt} = \frac{P_{①opt}}{\sigma_y} = \frac{P}{2\sigma_y \cos\left(\dfrac{90-\phi}{2}\right)} \qquad (5\text{-}14)$$

An important consideration in any systems study is sensitivity. Although a single value of θ is optimum for given ϕ, we may find that total weight is relatively insensitive to θ in the vicinity of θ_{opt}.* For example, if we were to establish a 5 percent weight penalty limitation for a minimum weight configuration we would find that a considerably large range of θ's about the optimum meet this requirement. This is illustrated in Fig. 5-5 where, for $\phi = 20°$, $f_e(\theta)$ has been plotted as a function of θ. Although as predicted by Eq. (5-13), $\theta_{opt} = 35°$, we see that for an allowable weight penalty of 5 percent any θ in the range $25° < \theta < 45°$ will suffice. Sensitivity studies are important since they provide the designer with flexibility in determining an acceptable or near optimum configuration by establishing the immediate effects of off-optimum selections on the merit function.

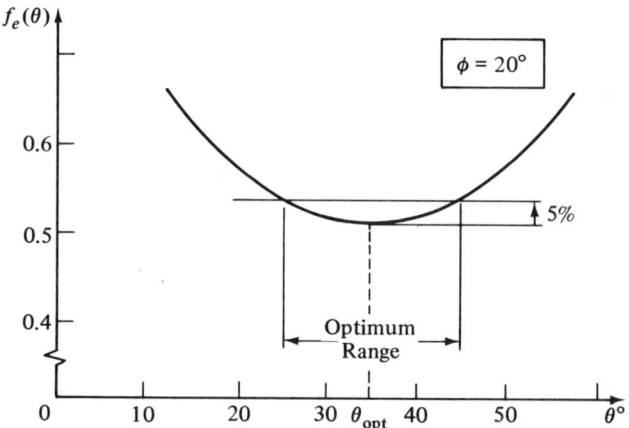

Fig. 5-5. Weight Sensitivity Evaluation for Two-cable System

As a second example of determinate similar-element systems consider the two beam system of Fig. 5-6 for the given environment, $E_s = \{q, L, D\}$. The specification of form will be a "T" configuration of uniform H-section simply supported beams with fixed orientation.

---

*A lack of weight sensitivity in the vicinity of $\{S_o\}_{opt}$ is fundamentally inherent to systems design. This generalization can be useful in a numerical evaluation of $\{S_o\}_{opt}$ since it justifies an iterative search with moderate, as opposed to refined, increments on each S_o.

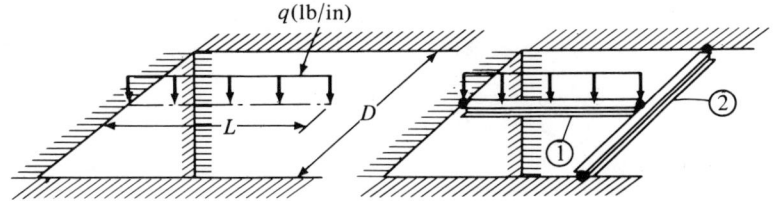

Fig. 5-6. Two-beam System with Fixed Orientation

Since the system admits to no orientation variables (referred to as *fixed orientation*) and is determinate, the local environments (maximum bending moments) can be evaluated in terms of $\{E_s\}$ only, namely: for the beam of span L and uniform load distribution q, the maximum bending moment occurs at the center span and can be expressed for simple end supports as

$$M_① = \frac{qL^2}{8} \tag{5-15}$$

For the beam of span D and center span load $qL/2$, the maximum bending moment also occurs at the center span and can be expressed for simple end supports as

$$M_② = \frac{qLD}{8} \tag{5-16}$$

For an *H*-section beam, we know from Eq. (4-37) that*

$$A_{\text{opt}} = \frac{1.606 M^{2/3}}{E^{1/6} \sigma_{PL}^{1/2}} \tag{5-17}$$

Assuming the same material for both beams, we can express the total system weight as

$$W_s = \rho A_① L + \rho A_② D = \frac{1.606 \rho}{E^{1/6} \sigma_{PL}^{1/2}} [M_①^{2/3} L + M_②^{2/3} D] \tag{5-18}$$

Now combining Eqs. (5-15), (5-16), and (5-18) yields

$$W_s = \frac{1.606 \rho q^{2/3} L^{7/3}}{(8)^{2/3} E^{1/6} \sigma_{PL}^{1/2}} \left[1 + \left(\frac{D}{L}\right)^{5/3} \right] \tag{5-19}$$

which completes the evaluation since the expression contains only environment and material.

Before leaving this example consider the following parametric question: If $D > L$ would a single-element cantilever as shown in Fig. 5-7 represent a weight savings over the simply supported beam system considered above?

---

*In this and subsequent applications we will employ the "better-proportioned" 2 percent compromise cross section of Eq. (4-37).

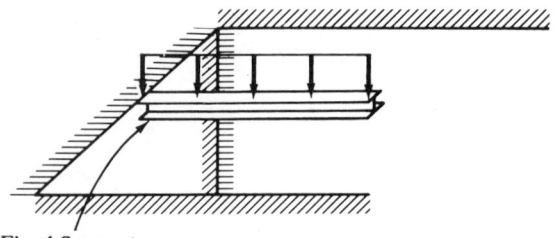

Fixed Support

Fig. 5-7. Single-element Cantilever Alternative

Certainly as D exceeds L without bound, a point must be reached where the presence of the support cross-member (element ②) represents diminished returns. To establish the limitation on D/L such that the system is optimum compared to the single-element cantilever, we first evaluate the cantilever weight as follows: the maximum moment occurs at the fixed end and can be expressed as

$$M_c = \frac{qL^2}{2} \tag{5-20}$$

Therefore, by Eqs. (4-37) and (5-20) we find

$$W_c = \rho A_c L = \frac{1.606 \rho q^{2/3} L^{7/3}}{(2)^{2/3} E^{1/6} \sigma_{PL}^{1/2}} \tag{5-21}$$

Now dividing through by L^3 in both Eqs. (5-19) and (5-21) yields

for the two-beam system:

$$\frac{W_s}{L^3} = \frac{1.606 \rho}{(8)^{2/3} E^{1/6} \sigma_{PL}^{1/2}} \left[1 + \left(\frac{D}{L}\right)^{5/3}\right]\left(\frac{q}{L}\right)^{2/3} \tag{5-22}$$

for the single-element cantilever:

$$\frac{W_c}{L^3} = \frac{1.606 \rho}{(2)^{2/3} E^{1/6} \sigma_{PL}^{1/2}} \left(\frac{q}{L}\right)^{2/3} \tag{5-23}$$

We see, therefore, that independent of material selection (E, σ_{PL}) and load index (q/L) the system will be more efficient than the cantilever alternative when

$$\frac{1}{(8)^{2/3}}\left[1 + \left(\frac{D}{L}\right)^{5/3}\right] < \frac{1}{(2)^{2/3}}$$

from which it is found that $D/L < 1.286$, for the requirement of $W_s < W_c$. Hence if D exceeds $1.286\,L$ the single-element cantilever will be more efficient.

It is common for the comparison of alternative systems to depend on some aspect of the environment. If we optimize systems parametrically in terms of environment, a comparison, such as illustrated above, can establish the relative efficiency between alternatives in these terms.

It should be noted that similar-element systems do not usually compete with single-element alternatives for conventional load environments. Conventional structural systems are typically composed of dissimilar elements which rely on a combination of load-carrying functions to yield a potentially more efficient structure than a single-element alternative. The examples considered in this section are isolated and were presented more as an illustration of the approach as opposed to fundamental applications.

Section 6
DETERMINATE DISSIMILAR-ELEMENT SYSTEMS

Consider the environment and two-bar truss system shown in Fig. 5-1. For the tension member Eq. (5-7) applies as follows

$$W_{①_{opt}} = \frac{\rho}{\sigma_y} P_① L_① \tag{5-24}$$

Specifying a circular tube column ($S_p = D, t$) for the compression member, its weight can be expressed as

$$W_{②_{opt}} = \frac{\rho}{\sigma_{A_{opt}}} P_② L_② \tag{5-25}$$

where from Eq. (2-20), (assuming pinned ends)

$$\sigma_{A_{opt}} = 0.54 \eta_T^{1/2} E^{2/3} \left(\frac{P_②}{L_②^2}\right)^{1/3} \tag{5-26}$$

$$\sigma_{A_{opt}} \leq \sigma_y$$

The local environments are as shown in Fig. 5-1. Substituting these values into Eqs. (5-24), (5-25), and (5-26) and summing Eqs. (5-24) and (5-25) yields for total system weight*

$$\frac{W_s}{L^3} = \left\{ \frac{\rho}{\sigma_y}\left[\frac{\cos\theta_2}{\cos\theta_1 \sin(\theta_1+\theta_2)}\right] + \frac{\rho}{\sigma_{A_{opt}}}\left[\frac{\cos\theta_1}{\cos\theta_2 \sin(\theta_1+\theta_2)}\right]\right\}\left(\frac{P}{L^2}\right) \tag{5-27}$$

$$\sigma_{A_{opt}} = 0.54 \eta_T^{1/2} E^{2/3} \left[\frac{\cos\theta_1 \cos^2\theta_2}{\sin(\theta_1+\theta_2)}\right]^{1/3} \left(\frac{P}{L^2}\right)^{1/3} \tag{5-28}$$

$$\sigma_{A_{opt}} \leq \sigma_y$$

where, in Eq. (5-27), we have divided through by L^3 to generate an index form.

Note from Eq. (5-27) that for a fully stressed column ($\sigma_{A_{opt}} = \sigma_y$) the system reduces mathematically to a similar-element system in that individual

---
*Assuming like material for both elements. We shall employ the convention of not subscripting material variables to indicate a common material selection for all elements

merit functions exhibit identical dependence on material and fixed-system environment. For this limiting case we can rewrite Eq. (5-27) as

$$\frac{W_s}{L^3} = \underbrace{\left\{\frac{1}{\sin(\theta_1 + \theta_2)}\left[\frac{\cos\theta_2}{\cos\theta_1} + \frac{\cos\theta_1}{\cos\theta_2}\right]\right\}}_{f_e(\theta_1, \theta_2)} \frac{p}{\sigma_y}\left(\frac{P}{L^2}\right) \quad (5\text{-}29)$$

for $\sigma_{A\mathrm{opt}} = \sigma_y$. Hence θ_1 and θ_2 can be optimized independent of P/L^2, σ_y, and p under the proviso that the numerical values of P/L^2, and σ_y are such that the column is fully stressed at optimum conditions.

We can take advantage of the product form of $f_e(\theta_1, \theta_2)$ in Eq. (5-29) by evaluating a stationary W_s using

$$\frac{d}{d(\theta_1 + \theta_2)}\left(\frac{1}{\sin(\theta_1 + \theta_2)}\right) = 0 \quad (5\text{-}30)$$

$$\frac{d}{d\theta_2}\left(\frac{\cos\theta_2}{\cos\theta_1} + \frac{\cos\theta_1}{\cos\theta_2}\right) = 0 \quad (5\text{-}31)$$

Eq. (5-30) is satisfied when

$$\theta_{1\mathrm{opt}} + \theta_{2\mathrm{opt}} = 90° \quad (5\text{-}32)$$

and from Eq. (5-31) we find

$$\frac{\cos^2\theta_1}{\cos^2\theta_2} = 1$$

from which

$$\cos\theta_1 = \cos\theta_2$$

or

$$\theta_{1\mathrm{opt}} = \theta_{2\mathrm{opt}} \quad (5\text{-}33)$$

Therefore by Eqs. (5-32) and (5-33) we conclude that

$$\theta_{1\mathrm{opt}} = \theta_{2\mathrm{opt}} = 45° \quad (5\text{-}34)$$

at a fully stressed optimum.

From Eqs. (5-27) and (5-28) it appears that in order to obtain values of $\theta_{1\mathrm{opt}}$ and $\theta_{2\mathrm{opt}}$ for an understressed optimum column, both material and environment must be numerically specified. Although this is true for the individual material variables (E, σ_y), we have lumped the environment (P, L) into a load index (P/L^2) and can therefore evaluate in terms of this index for given materials variables. By combining Eq. (5-27) with Eq. (5-28) we find

$$\frac{W_s}{L^3} = p\left[\frac{\cos\theta_2\,(P/L^2)}{\sigma_y \cos\theta_1 \sin(\theta_1 + \theta_2)}\right] + p\left[\frac{\cos^{2/3}\theta_1\,(P/L^2)^{2/3}}{0.54\eta_T^{1/2}E^{2/3}\cos^{5/3}\theta_2 \sin^{2/3}(\theta_1 + \theta_2)}\right] \quad (5\text{-}35)$$

for $\sigma_{A\mathrm{opt}} < \sigma_y$. Note from Eq. (5-35) that for each selection of P/L^2, optimum θ's can be numerically evaluated and the corresponding minimum value of

W_s/L^3 (a weight index) can thus be determined in terms of P/L^2. This numerical solution to Eq. (5-35) is shown in Fig. 5-8 for AISI 1025 steel.

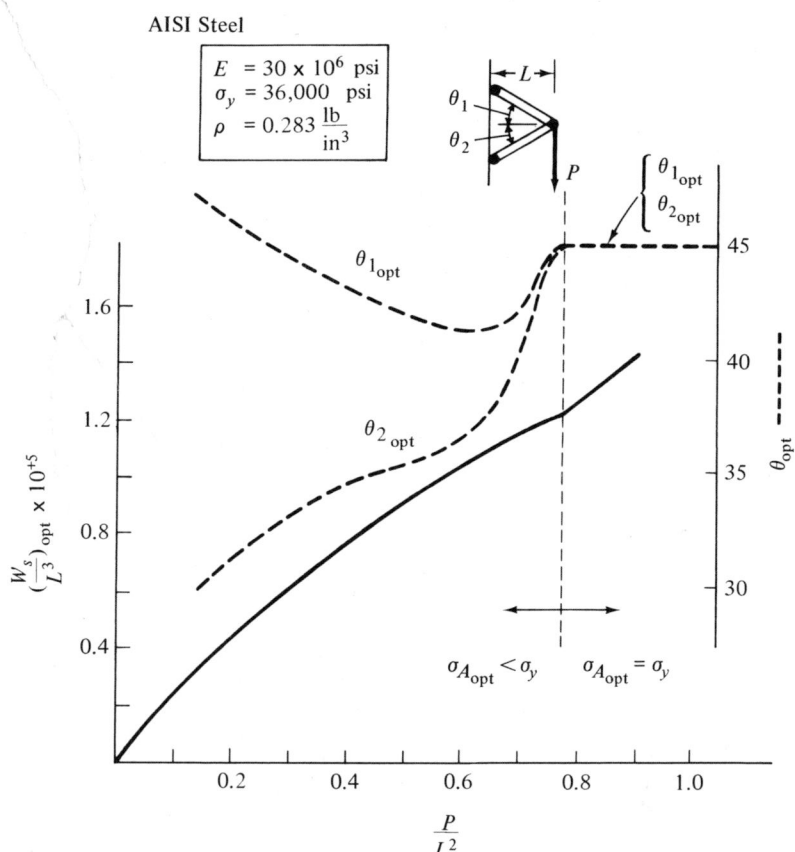

Fig. 5-8. Optimum Orientation and Weight Index in Terms of P/L^2

Figure 5-8 represents a complete solution to the two-bar truss for the given material.* For any values of P and L the above curves will determine $W_{s_{opt}}$, $\theta_{1_{opt}}$, and $\theta_{2_{opt}}$. With these values known, local environments follow from the equations of Fig. 5-1, and, therefore, optimum proportions will follow from the single-element relationships developed in previous chapters.

*Note that the sum $\theta_{1_{opt}} + \theta_{2_{opt}}$ is not equal to 90° in the region $\sigma_{A_{opt}} < \sigma_y$. For example at $P/L^2 = 0.3$, $\theta_{1_{opt}} = 44°$, $\theta_{2_{opt}} = 33°$ and therefore $\theta_1 + \theta_2 = 77°$. The tendency for θ_2 to be less than θ_1 in this region indicates that at the understressed optimum, column buckling effects require a reduced column length compared to the tension member.

The preceding design solution for the two-bar truss can be considered representative of more complex truss networks. By definition a truss network is a combination of straight elements mutually pin-connected at the ends with external restraints and loads at pin locations only. Inherent in this definition a truss will be composed of tension and compression members only. As such, the design of a more involved multiple-element truss form has the effect of changing the number of terms and the number and functional form of orientation variable angles in Eq. (5-35). The index form of Eq. (5-35) and the manner in which it was constructed can be considered a generalized approach for trusses.*

The indicized relationship of Eq. (5-35), W_s/L^3 as a function of P/L^2, is not peculiar to tension or compression members but inherent to the environment of (P, L). To further demonstrate the versatility of this manner of displaying results for the given environment, consider a single uniform H cross-section cantilever as an alternative to the two-bar truss of Fig. 5-1. The maximum moment occurs at the wall and is equal to PL. Therefore, by Eq. (4-37)

$$A_{B_{opt}} = \frac{1.606}{E^{1/6}\sigma_{PL}^{1/2}}(PL)^{2/3} \qquad (5\text{-}36)$$

The weight can then be expressed as

$$W_B = \rho AL = \frac{1.606\rho}{E^{1/6}\sigma_{PL}^{1/2}}(PL)^{2/3}L \qquad (5\text{-}37)$$

Now, dividing by L^3 we find

$$\frac{W_B}{L^3} = \rho \frac{1.606}{E^{1/6}\sigma_{PL}^{1/2}}\left(\frac{P}{L^2}\right)^{2/3} \qquad (5\text{-}38)$$

which results in a similar indicized relation for the cantilever alternative. For a given material we can compare both forms for the general environment as shown in Fig. 5-9.

From Fig. 5-9 it can be seen that the two-bar truss is more efficient than the uniform cantilever beam over a considerable range of load index. For example at $P/L^2 = 1.0$ the weight ratio is approximately 8. Eventually, due to the linear rise of truss weight at the fully stressed condition, a point must be reached where the beam is more efficient.

Systems Load Index

The load index comparison of alternative forms for the general environment is one of the most powerful subterfuges in the arsenal of approaches to the optimization of structures. We have seen previously in Chap. 2 and Chap. 4

---

*In Chap. 6 the systems approach for trusses will be generalized to include redundancy and multiple (nonconcurrent in time) load conditions.

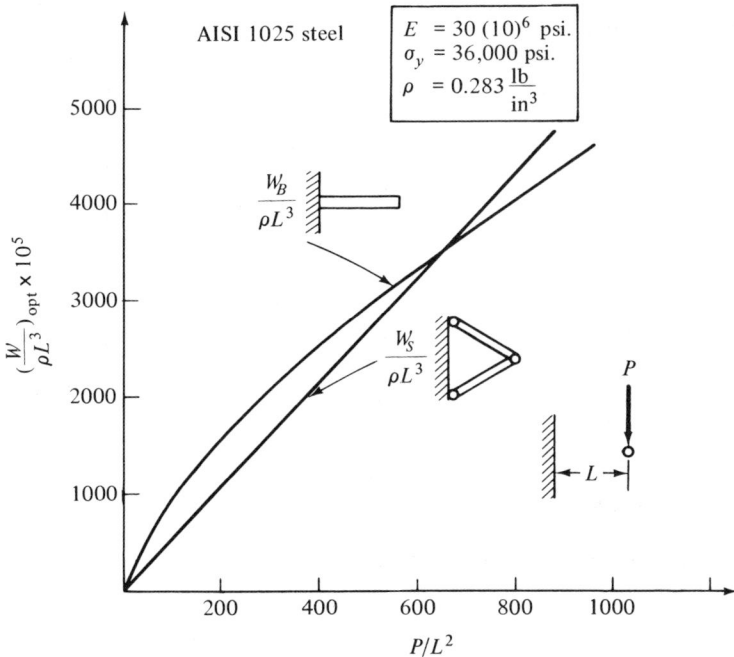

Fig. 5-9. Comparison of Two-bar Truss and Cantilever for the General Environment

how a load index formulation can facilitate alternative form comparisons for a single element. We now see the potential of extending this approach to systems involving a combination of similar or dissimilar elements. Before considering the indicized formulation of weight results under more complex system environments, let us pause and explore the notion of load index, partly in retrospect but for the most part with aspirations to a more formal index approach for the general structural system.

Certainly the lumping of environmental parameters into a single common load index is desirable for form comparison. What concerns us then is how this index formulation is achieved for the general design problem. Shanley[6] has suggested that the load index will always have the units of stress. For design problems involving a load P transmitted through a distance L—whether L is collinear to P (the column problem) or perpendicular to P (the truss or beam problem)—the appropriate load index has been shown to be P/L^2. We have also seen that a distributed load q (in lb/in) over a beam of span L yields a load index of q/L, again in psi. The general wide column of length L under distributed load q can also be shown to yield a load index of q/L (Ref. 13). In general, in any system environment which is manifest in terms of a single load and a single transmission path, the proper ratio of the load agent to the length agent (in units of stress) can be considered an inherent load index.

Such a "stress generalization" is useful in reducing a particular weight result to index form. As an a priori tool, however, it is limited in that if the environment exhibits more than one transmission length such as illustrated in Fig. 5-10, no basis exists for deciding between the possible indices prior to considering the constructed weight expressions for alternative forms.*

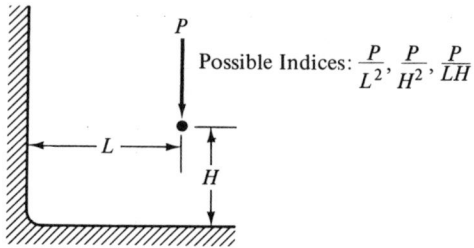

Fig. 5-10. Problem of Determining Load Index Prior to Form Specification and Optimization

The desired form of a weight-index/load-index relationship, if not inherent to the nature of the environment, will depend on the nature of alternative form specifications. For our purposes, therefore, we will be content with utilizing the stress generalization as an aid in determining the load index based on a "comparison requirement" between alternative forms. The following example will illustrate this aspect of systems design.

Consider the environment of Fig. 5-11(a). A uniformly distributed load is to be carried across a span L at a height H. A basic single element design solution would be a simple beam as shown in Fig. 5-11(b).

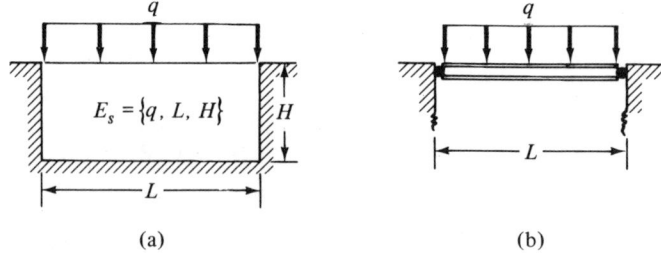

Fig. 5-11. Environment and Basic Single-element Design Solution

*See Prob. 5-5.

If the ratio of H/L is small it seems apparent that column supports with optimized orientation in combination with a beam represents a system which should compete with the beam alone. In what follows we shall explore this fundamental trade-off in terms of the general environment by reducing our weight results to index form.

First, for the beam alone, we shall specify a uniform H cross section in which case Eq. (4-37) applies for A_{opt}. The maximum moment is simply $qL^2/8$. Substituting this moment into Eq. (4-37) and multiplying ρA_{opt} by L yields for beam weight

$$W_B = \rho \frac{1.606 q^{2/3} L^{7/3}}{(8)^{2/3} E^{1/6} \sigma_{PL}^{1/2}} \tag{5-39}$$

Now dividing through by L^3 yields

$$\frac{W_B}{L^3} = \rho \frac{1.606}{(8)^{2/3} E^{1/6} \sigma_{PL}^{1/2}} \left(\frac{q}{L}\right)^{2/3} \tag{5-40}$$

as one possible index result.*

Consider now the system alternative employing a combination beam-column bridge as shown in Fig. 5-12. We shall specify a uniform H cross section beam and circular tube columns where the column spacing (b) is an orientation variable.

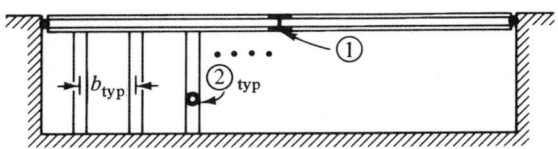

Fig. 5-12. Beam–Column Bridge System

For a small number of column supports this system must be considered as N-redundant where N is the number of columns.† For long spans ($L \gg H$) the optimum number of column supports will likely be large compared to the two end supports. Making this provisional assumption we shall neglect the portion of the load carried by the end supports. Hence by symmetry we can evaluate the load transmitted to each column as

$$P_{②} = qb \tag{5-41}$$

Also we recognize that

$$L_{②} = H \tag{5-42}$$

---

*We shall see presently that an alternative index form is possible. This alternate form is suggested by the comparison requirement which follows.

†We shall consider this problem in Chap. 6.

which completes the recognition of column local environment. For the local environment of the beam it is found that the maximum bending moment occurs at the column supports as shown in Fig. 5-13 (assuming stationary pin supports).

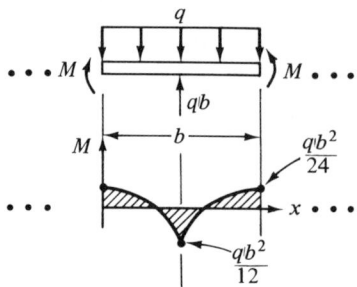

Fig. 5-13. Local Environment for the Beam

Whence

$$M_① = \frac{qb^2}{12} \tag{5-43}$$

For a single beam and N columns the total system weight can be written as

$$W_s = \rho A_① L + N\rho A_② H \tag{5-44}$$

By combining Eqs. (4-37) and (5-43) we find, for $A_①$,

$$A_① = \frac{1.606 q^{2/3} b^{4/3}}{(12)^{2/3} E^{1/6} \sigma_{PL}^{1/2}} \tag{5-45}$$

The area expression for a column can be written as

$$A_② = \frac{P_②}{\sigma_{A_{\text{opt}}}} \tag{5-46}$$

which, when combined with Eq. (5-41), yields

$$A_② = \frac{qb}{\sigma_{A_{\text{opt}}}} \tag{5-47}$$

Now combining Eqs. (5-44), (5-45), and (5-47) results in

$$W_s = \rho \frac{1.606 q^{2/3} b^{4/3} L}{(12)^{2/3} E^{1/6} \sigma_{PL}^{1/2}} + \rho \frac{qHL}{\sigma_{A_{\text{opt}}}} \tag{5-48}$$

where, under the assumption of neglecting end effects, we have expressed the number of columns, N, as $N = (L/b)$. Also, by Eqs. (2-20), (5-41), and (5-42)

$$\sigma_{A_{\text{opt}}} = 0.54 \eta_T^{1/2} E^{2/3} \left(\frac{qb}{H^2}\right)^{1/3} \tag{5-49}$$

$$\sigma_{A_{\text{opt}}} \leq \sigma_y$$

Since the system environment and system weight equations are expressed in terms of two transmission paths (L and H) there is no inherent load index. We shall see in the following evaluation and Prob. 5-6 that either q/L or q/H will result in a weight-index/load-index formulation of Eqs. (5-48) and (5-49). If we were to be guided by the seemingly inherent index formulation obtained for the basic single beam comparator (Eq. (5-40)), we would select q/L. Since Eqs. (5-48) and (5-49) represent a more involved weight relationship, we will choose to be guided by the index formulation which is most easily obtained for them, taking the attitude that we can reconstruct Eq. (5-40) to match this formulation. This is recommended as a general attitude when an indicized comparison is sought between alternatives of differing complexity.

Proceeding on this basis we note that q/H represents a convenient index for Eq. (5-49) and yields

$$\sigma_{A_{opt}} = 0.54 \eta_T^{1/2} E^{2/3} \left(\frac{b}{H}\right)^{1/3} \left(\frac{q}{H}\right)^{1/3} \tag{5-50}$$

where b/H can be considered an *orientation index*. We then find that Eq. (5-48) can be converted to this index form by dividing both sides by LH^2 which results in

$$\frac{W_s}{LH^2} = \rho \left[\frac{1.606}{(12)^{2/3} E^{1/6} \sigma_{PL}^{1/2}} \left(\frac{b}{H}\right)^{4/3} \left(\frac{q}{H}\right)^{2/3} + \frac{\left(\frac{q}{H}\right)}{\sigma_{A_{opt}}}\right] \tag{5-51}$$

To reconstruct Eq. (5-40) to match this form, we first rewrite to obtain a q/H load index as shown

$$\frac{W_B}{L^3} = \rho \frac{1.606}{(8)^{2/3} E^{1/6} \sigma_{PL}^{1/2}} \left(\frac{q}{H}\right)^{2/3} \left(\frac{H}{L}\right)^{2/3} \tag{5-52}$$

This necessitates the introduction of the ratio H/L, which should be no surprise since the comparison of these alternatives must depend on this ratio.* Dividing both sides of Eq. (5-52) by $(H/L)^2$ completes the reconstruction since W_B/L^3 is thus converted to W_B/LH^2, viz.

$$\frac{W_B}{LH^2} = \rho \frac{1.606}{(8)^{2/3} E^{1/6} \sigma_{PL}^{1/2}} \frac{\left(\frac{q}{H}\right)^{2/3}}{\left(\frac{H}{L}\right)^{4/3}} \tag{5-53}$$

We now have common indicized relations for both forms. Before considering the graphical trade-off between these alternatives in terms of H/L, we can first determine an analytical optimum for b/H for the design region where

---

*It should be apparent now why q/H was selected as the load index. By this selection, the parameter H/L results in Eq. (5-52) (the more simple relation) as opposed to Eq. (5-51) (see Prob. 5-6).

columns are understressed, i.e. ($\sigma_{A_{opt}} < \sigma_y$). To obtain this result we combine Eqs. (5-50) and (5-51), which yield

$$\frac{W_s}{LH^2} = \rho \left[\frac{1.606}{(12)^{2/3} E^{1/6} \sigma_{PL}^{1/2}} \left(\frac{b}{H}\right)^{4/3} + \frac{1}{0.54 \eta_T^{1/2} E^{2/3} \left(\frac{b}{H}\right)^{1/3}} \right] \left(\frac{q}{H}\right)^{2/3} \qquad (5\text{-}54)$$

Since $(q/H)^{2/3}$ factors we see that the orientation index, b/H, can be optimized with dependence only on material variables. Differentiating Eq. (5-54) with respect to b/H and equating to zero results in

$$\left(\frac{b}{H}\right)_{opt} = \left[\frac{(12)^{2/3}}{4(.54)(1.606)}\right]^{3/5} \left(\frac{\sigma_{PL}}{E}\right)^{3/10} \qquad (5\text{-}55)$$

For example, employing AISI 1025 steel Eq. (5-55) yields $(b/H)_{opt} = 0.17$. Fig. 5-14 illustrates the comparison between the two forms for this material selection in the design region, $\sigma_{A_{opt}} < \sigma_y$.

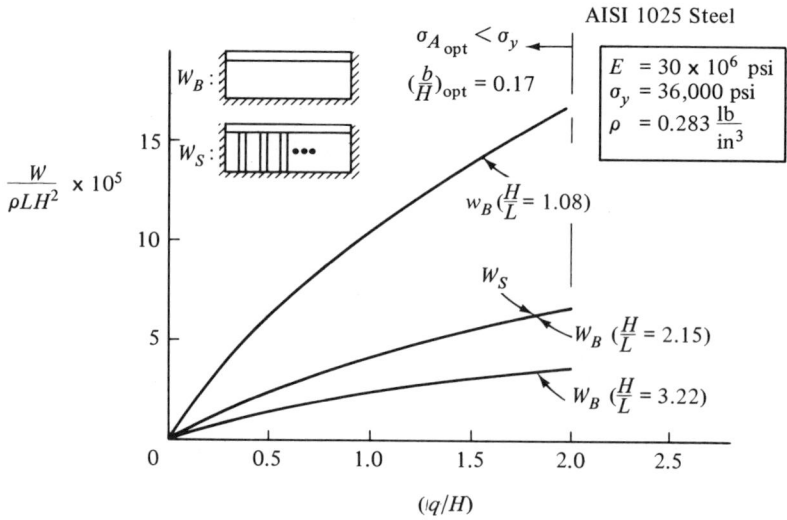

Fig. 5-14. Comparison of Beam–Column Bridge and Single Beam

Note that for $H/L = 2.15$ the single-beam weight and beam-column system weight become coincident. For $H/L < 2.15$ the system is seen to be more efficient for the general environment.

Section 7
GEOMETRIC CONSTRAINTS

In Chaps. 2 and 4 the technique of employing parametric variation of slack variables to effect geometric constraints for single elements was illustrated. In Chap. 7 a similar manipulation of excess proportion variables—as opposed to slack variables—will be considered as a mechanism for achieving such constraints with a minimized merit penalty.

In a systems evaluation recognize that all proportion variables are dependent evaluations for any specified values of the orientation and redundant load variables. As such if component element relations can be modified to include geometric constraints they can be incorporated readily into a system study. This would be accomplished simply by checking and modifying each geometrically constrained proportion variable for each trial design point $\{S_o, R_N\}$.

Section 8
MULTIPLE LOAD CONDITIONS

The systems approach discussed in this chapter can be readily modified to include multiple (nonconcurrent in time) loading conditions. For multiple load cases Eq. 5-1 must be replaced by

$$E_{si} = \text{the extreme of} \begin{cases} E_{si,1}(E_S, S_o, R_N) \\ E_{si,2}(E_S, S_o, R_N) \\ \vdots \\ E_{si,K}(E_S, S_o, R_N) \end{cases} \quad (5\text{-}56)$$

where K is the number of potentially extreme load cases for a given element. Note that in this way the merit function (Eq. (5-3)) is defined piece wise since the functional form of each E_{si} will depend on the current design point $\{S_o, R_N\}$.

Applications under multiple load conditions will be considered in Chap. 6, Sec. 6.

PROBLEMS

5-1. Show that for $\phi = 90°$ the weight of the two-cable system of Fig. 4-6 (b) reduces to a single vertical cable configuration at optimum conditions.

5-2. Show that the following twin cantilever system is of minimum weight at $\theta = 0$.

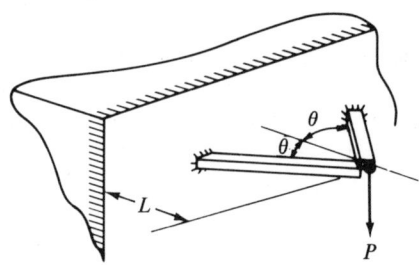

Compare this optimized twin cantilever to a single cantilever alternative and show that it is less efficient. Make the comparison with indicized weight relations.

5-3. Develop optimum weight expressions in the form of Eqs. (5-27) and (5-28) for the truss shown below.

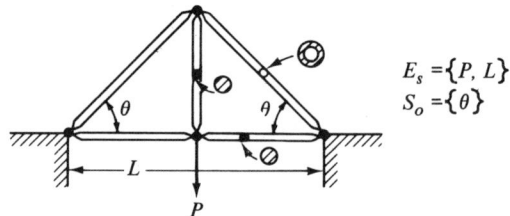

Show that this yields

$$\frac{W_s}{L^3} = \left(\frac{P}{\sigma_y}\frac{\cot\theta}{2} + \frac{P}{\sigma_y}\frac{\tan\theta}{2} + \frac{P}{\sigma_{A_{\text{opt}}}}\frac{1}{\sin 2\theta}\right)\left(\frac{P}{L^2}\right)$$

where

$$\sigma_{A_{\text{opt}}} = 0.54\eta_T^{1/2}E^{2/3}\left(\frac{2\cos^2\theta}{\sin\theta}\right)^{1/3}\left(\frac{P}{L^2}\right)^{1/3}, \quad \sigma_A < \sigma_y$$

Show that for fully stressed columns ($\sigma_{A_{\text{opt}}} = \sigma_y$), $\theta_{\text{opt}} = 45°$. Evaluate the understressed condition numerically for AISI 1025 steel. Plot $(W_s/L^3)_{\text{opt}}$ and θ_{opt} in terms of (P/L^2) for this material.

5-4. Superimpose the plot $(W_B/L^3)_{\text{opt}}$ in terms of (P/L^2) for a single H-section beam as an alternative for the environment and form of Prob. 5-3. At what value of (P/L^2) is the beam more efficient?

5-5. Show that weight of the alternative forms below can be compared on the index basis of W/L^3 in terms of P/L^2 or W/H^3 in terms of P/H^2.

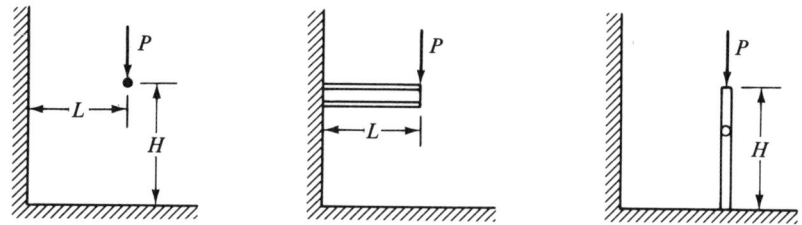

From either of the above evaluations determine a relationship for H/L such that the beam is more efficient.

5-6. Reconstruct the weight expression for the beam column system (Eq. 5-54) on a W_s/L^3 in terms of q/L basis. Show that this yields

$$\frac{W_s}{L^3} = \rho \left[\frac{1.606}{(12)^{2/3} E^{1/6} \sigma_{LP}^{1/2}} \left(\frac{b}{H}\right)^{4/3} \left(\frac{H}{L}\right)^{4/3} + \frac{\left(\frac{H}{L}\right)^{4/3}}{0.54 \eta_T^{1/2} E^{2/3} \left(\frac{b}{H}\right)^{1/3}} \right] \left(\frac{q}{L}\right)^{2/3}$$

Note that with this expression the beam column system can be compared with the single-beam alternative as given in Eq. (5-40). Why would this be less desirable than the comparison made in Fig. 5-14?
Hint: The system weight expression of Eq. (5-54) does not contain the parameter H/L.

5-7. For the environment shown $E_s = \{q, L, D, \ell\}$, specify a T-configuration of H-section simply supported beams with variable orientation b.

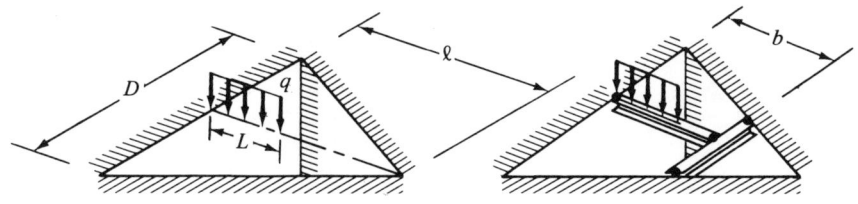

Show that system weight can be expressed as

$$\frac{W_s}{L^3} = \frac{1.606 \rho}{E^{1/6} \sigma_{PL}^{1/2}} \left[\frac{(b-L)^2}{(2)^{2/3} L^{5/3} b^{1/3}} + \frac{(\ell - b)^{2/3} D^{5/3}}{(8)^{2/3} L b^{2/3} \ell^{2/3}} \right] \left(\frac{q}{L}\right)^{2/3}$$

5-8. For the environment below a two-bar truss must be designed with constraints on θ_1 and θ_2 as shown.

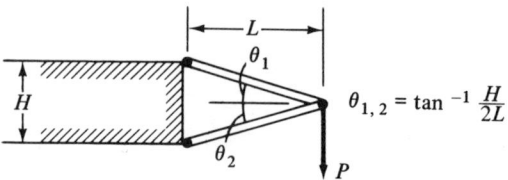

When H is less than L this can yield large off-optimum weight penalties (the constrained small values of θ_1 and θ_2 result in large local loads in order to substain P). To avoid these penalties the truss form below is proposed.

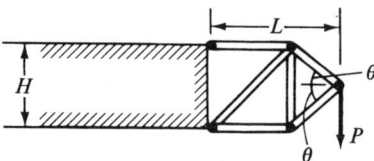

Evaluate the system weight for this alternative in index form. Show that at a fully stressed optimum, θ_{opt} is dependent only on H/L. Compare at this condition with a two-bar truss ($H/L = 0.5$) and calculate the weight ratio.

5-9. At what value of H/L does the truss below become more efficient than that considered in Prob. 5-8 (evaluate at the fully stressed condition)?

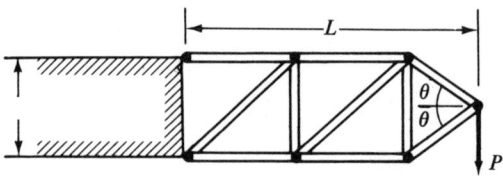

5-10. Find a relationship for H/L for which a uniform H-section cantilever becomes superior to the two-bar truss of Prob. 5-8.

5-11. Compare a simply supported H-section beam to the truss of Prob. 5-3. Make the evaluation for Al 7075-T6 and a fully stressed design for the truss. Determine the range of P/L^2 for which the truss is more efficient.

6

DESIGN OF STATICALLY INDETERMINATE STRUCTURES

In Chap. 5 a systems approach to structural design was developed based on the ability to evaluate local environments for each component element in terms of the fixed-system environment, orientation variables, and redundant loads. The applications considered in Chap. 5, however, were statically determinate systems. In this chapter we extend these applications to redundant structures in order to illustrate the ability to consider redundant loads as independent design variables and additionally, to draw some general conclusions regarding the potential of redundancy for the improvement of the weight efficiency of certain structural combinations.

The approach utilizes equilibrium equations to evaluate local loads in terms of the redundant loads. In this way merit can then be expressed in these terms with optimized redundant loads following from a statement of maximized merit. Since redundant elements always represent an excess over and above the basic static configuration, they can be adjusted in the realized structure so as to carry the optimized state of load while satisfying compatibility of deformation. (This will be discussed in detail in Sec. 4.)

Section 1

REDUNDANT EXTERNAL RESTRAINT

The use of excess external restraint, such as restraint against rotation at the ends in a beam or column as opposed to pinned ends, will generally result in reduced structural weight. We have already seen in Chap. 2 that fixed ends in a column represent a 36 percent weight savings over the pinned condition. This was not a strict case of redundancy since in the stable configuration, column loads are statically determinate for both pinned and fixed ends. An element carrying transverse loads, on the other hand, develops local moment loads which vary with the degree of fastening restraint and thus here we find the potential of optimizing such restraint.

Certainly there may be limitations on the number and type of external restraints dependent on the nature of the existent material environment. Furthermore, there will be diminished weight returns in the form of increased fastening weight. Whether these returns are diminished to a point of no gain depends on the design detail of the fastening scheme and on the length of the element. The assessment of fastening weight is complicated by the fact that the design of fasteners has survived to the present as principally an art with very little in the way of analytical approach. Although the analysis of various types of fastening schemes employing bolts, rivets, welds, etc., has reached a level of sophistication in a blend of empirical and analytical procedures, the synthesis of fastening schemes has relied primarily on iterative techniques.

It is not our objective here to treat the design of fastening schemes using analytical optimization but rather to evaluate the optimum degree of restraint considering the structural weight of the element alone. Recognize that whenever the increased fastening weight is assessed, whether the additional restraint is justified or not will require analytical procedures to determine the structural weight of the member as a function of the amount of restraint.

As an example consider a uniformly distributed load to be carried by a uniform H-section beam over a span as shown in Fig. 6-1.

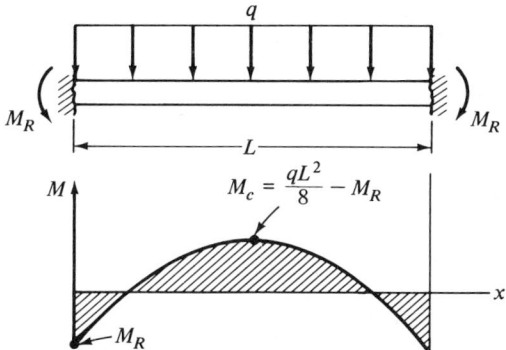

Fig. 6-1. Example of Redundant End Restraint

We seek to optimize the degree of end restraint by determining the optimum value of the redundant end moment, M_R. From the moment diagram we recognize that either the redundant moment or the center span moment, M_c, may become critical. For a uniform cross section the local environment must obviously be taken as the larger of these two moments. Since the problem is statically indeterminate, both these loads must be evaluated in terms of the redundant moment, M_R. To simplify the analysis we express the magnitude of M_R as

$$M_R = kqL^2 \tag{6-1}$$

thereby treating k as the redundant variable. From moment equilibrium, the center span moment can then be expressed in these terms as

$$M_c = \left(\frac{1}{8} - k\right)qL^2 \tag{6-2}$$

Employing Eqs. (6-1) and (6-2) the condition on k for which M_c exceeds M_R is

$$\left(\frac{1}{8} - k\right) > k$$

and therefore

$$k < \frac{1}{16} \tag{6-3}$$

becomes the condition for k such that $M_c > M_R$.

From Eq. (4-37), the uniform beams weight for the maximum local moment is

$$W = \rho L \frac{1.606}{E^{1/6}\sigma_{PL}^{1/2}} M_{\max}^{2/3} \tag{6-4}$$

Combining Eq. (6-4) with Eqs. (6-1) through (6-3) can be expressed as

$$W = \rho L \frac{1.606}{E^{1/6}\sigma_{PL}^{1/2}} (qL^2)^{2/3}\{f_e(k)\}^{2/3} \tag{6-5}$$

where

$$f_e(k) = \begin{cases} \left(\frac{1}{8} - k\right) & \text{for } k \leq \frac{1}{16} \\ k & \text{for } k \geq \frac{1}{16} \end{cases}$$

and gives the beam weight as a function of the redundant variable k.

Since for increasing k, $f_e(k)$ decreases when $k < 1/16$ and increases when $k < 1/16$, the optimum value of k is seen to be $1/16$ corresponding to equal magnitudes for M_R and M_c. This is illustrated graphically in Fig. 6-2 which shows $f_e(k)$ as a function of k.

It is interesting to note that the optimum degree of restraint is less than that achieved at the true fixed end (zero slope) condition. At the zero slope condition $M_R = 2M_c$, corresponding to $k = 1/12$ and a weight increment of 21 percent over the optimized condition of restraint. Note also that optimum restraint occurs when $M_R = M_c$. There is an evident generalization here which will become more apparent in subsequent sections: *Whenever, as the restraint increases, the pertinent local load increases at one point on the member and decreases at another, the optimum restraint will be when both loads are equal.*

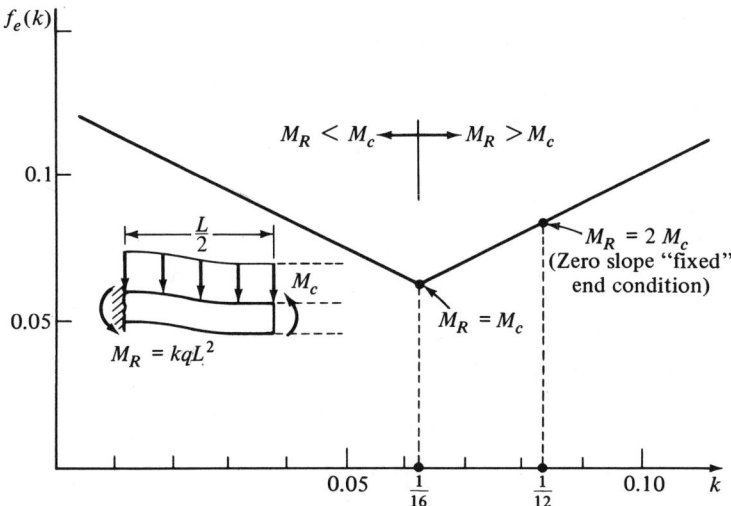

Fig. 6-2. Weight Efficiency as a Function of Redundant Restraint

Section 2
REDUNDANT SIMILAR-ELEMENT SYSTEMS

There is a potential for an optimum redundant combination of beam elements depending on the nature of the load environment. If the environment is such that a redundant element offers efficient support to the determinate systems but, by itself, cannot support the loads, the redundant combination will be optimum. If, on the other hand, a redundant element offers efficient support to the determinate systems but, by itself, or in determinate combination, could support the loads, the determinate configuration will be optimum. We shall illustrate this by the following two examples.

Consider first the environment of Fig. 6-3. A single load P is to be carried by a redundant combination of uniform beams cantilevered at the walls and pin-connected at the point of applied load as shown.

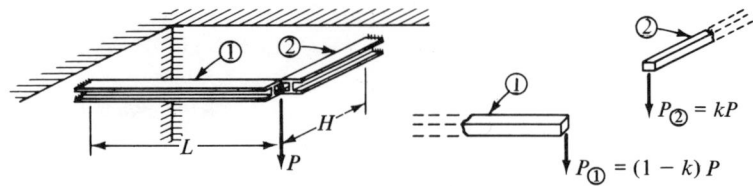

Fig. 6-3. Example of a Redundant Beam System Where Either Beam Alone Can Support Loads

As illustrated in the figure, we have taken $P_① = (1-k)P$ and $P_② = kP$ which defines k as a redundant variable indicative of the redundant force split between members ① and ②. The local environments (maximum moments) can then be expressed as

$$M_① = (1-k)PL \tag{6-6}$$
$$M_② = kPH$$

Employing Eq. (4-37) for the component weights of each element yields for total system weight

$$W_s = \rho \frac{1.606}{E^{1/6}\sigma_{PL}^{1/2}}\{[(1-k)PL]^{2/3}L + (kPH)^{2/3}H\} \tag{6-7}$$

Factoring $P^{2/3}L^{5/3}$ and then dividing through by L^3 results in

$$\frac{W_s}{L^3} = \rho \frac{1.606}{E^{1/6}\sigma_{PL}^{1/2}}\underbrace{\left[(1-k)^{2/3} + k^{2/3}\left(\frac{H}{L}\right)^{5/3}\right]}_{f_e}\left(\frac{P}{L^2}\right)^{2/3} \tag{6-8}$$

In this form the bracketed factor, $f_e(k, H/L)$, can be minimized parametrically with respect to H/L. Fig. 6-4 shows f_e plotted as a function of k for various ratios of H/L.

It can be seen that for $H/L < 1$, $k_{opt} = 1$ and demonstrates that a single beam with the shorter transmission path, H, would carry the entire load at minimum weight. For $H/L > 1$, $k_{opt} = 0$ and demonstrates that again the shorter length member (this time of length L) carries the entire load at minimum weight. At $H/L = 1$ both $k = 0$ and 1 yield the same minimum weight. For no values of the system environment does a redundant combination of two beams ($0 < k < 1$) become optimum. The conclusion is that since either beam could transmit the load, whichever could do it the more efficiently yields a determinate optimum.

Suppose, however, that the environment is such that a redundant element, acting alone, could not possibly transmit the loads and yet represents an efficient load path. In such cases a redundant combination will be optimum. As an example, consider the load environment and redundant beam system of Fig. 6-5.

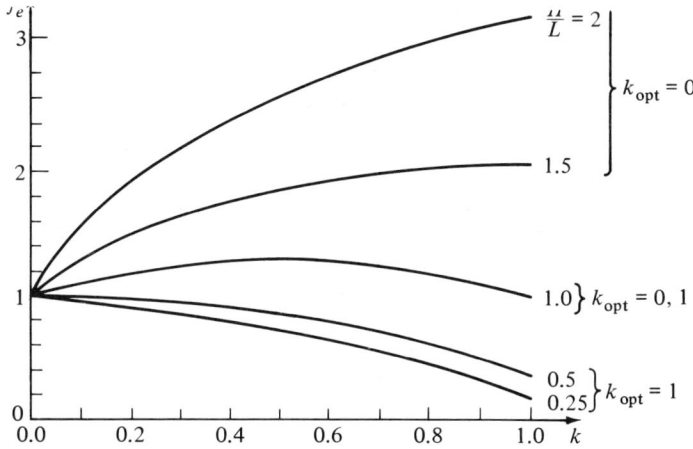

Fig. 6-4. Weight Efficiency as a Function of Redundant Force Split

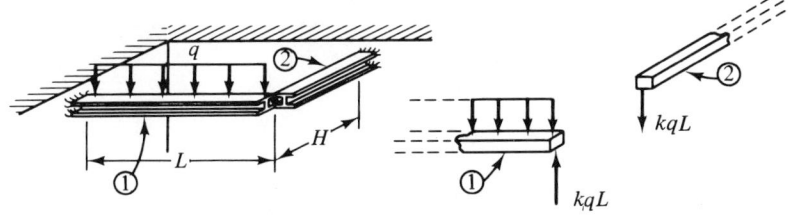

Fig. 6-5. Redundant Beam System Where Only One of the Beams Acting Alone Can Support the Loads

Expressing the redundant load transmitted to element ② as kqL, the local environment for this element becomes

$$M_{②} = kqLH \tag{6-9}$$

Determination of the critical local moment for element ① in terms of the redundant variable k requires a moment diagram evaluation as shown in Fig. 6-6 on p. 108. There are two moments which may become critical, one where the shear load vanishes, M_a, and one at the wall restraint, M_w. From the geometry of the shear and moment diagrams the *magnitudes* of these moments are found to be

$$M_a = \frac{1}{2}k^2qL^2$$

$$M_w = \left[\frac{1}{2}(1-k)^2 - \frac{1}{2}k^2\right]qL^2 \tag{6-10}$$

The condition on k for which say $M_a > M_w$ is thus

$$\frac{k^2}{2} > \frac{1}{2}(1-k)^2 - \frac{k^2}{2}$$

and yields

$$k > 0.415 \qquad (6\text{-}11)$$

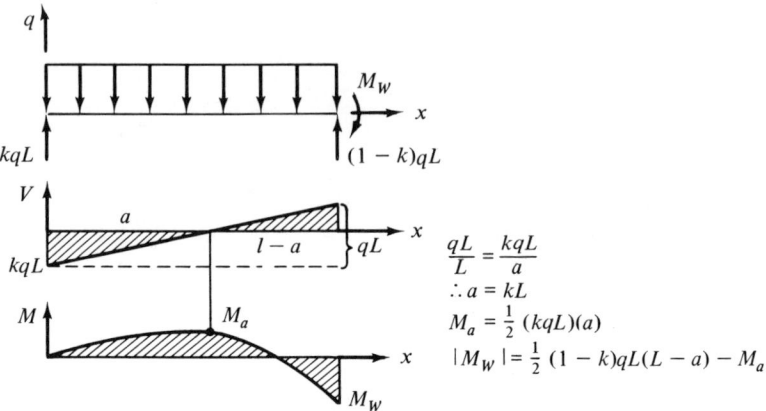

Fig. 6-6. Moment Evaluation for Element ①

Accordingly, application of Eq. (4-37) combined with Eqs. (6-9) and (6-10) yields for total system weight

$$W_s = \rho \frac{1.606}{E^{1/6}\sigma_{PL}^{1/2}}[(qL^2)^{2/3}\{f(k)\}^{2/3}L + k^{2/3}(qLH)^{2/3}H] \qquad (6\text{-}12)$$

where

$$f(k) = \begin{cases} \frac{1}{2}k^2 & \text{for } k \geq 0.415 \\ \frac{1}{2}(1-k)^2 - \frac{1}{2}k^2 & \text{for } k \leq 0.415 \end{cases}$$

Factoring $q^{2/3}L^{7/3}$ and dividing through by L^3 yields

$$\frac{W_s}{L^3} = \rho \frac{1.606}{E^{1/6}\sigma_{PL}^{1/2}}\underbrace{\left[\{f(k)\}^{2/3} + k^{2/3}\left(\frac{H}{L}\right)^{5/3}\right]}_{f_e}\left(\frac{q}{L}\right)^{2/3} \qquad (6\text{-}13)$$

As in the previous example, we can now determine the optimum value of k for which f_e attains a minimum parametrically with respect to H/L. This evaluation is shown in Fig. 6-7.

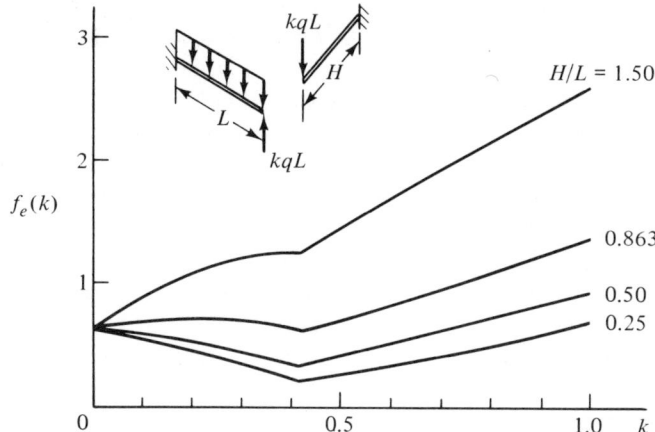

Fig. 6-7. Weight Efficiency as a Function of Redundant Variable k

Note that for any $H/L < 0.863$ a redundant combination ($k_{opt} = 0.415$) is optimum corresponding to 41.5 percent of the total applied load magnitude, qL, transmitted to element ②. At these lower ratios of H/L, element ② represents an efficient load path, but being unable to support the load alone, a redundant combination results at minimum weight. Note also that $k_{opt} = 0.415$ corresponds to equality of the competing moments M_a and M_w for element ①, a result typical of redundant systems.

Section 3
REDUNDANT DISSIMILAR-ELEMENT SYSTEMS

An excess of dissimilar elements offers a marked potential for providing efficient support to an otherwise determinate system. Such elements would be those required to provide the optimized amount of external restraint such as discussed in Sec. 1 and, more importantly, those elements that would provide efficient "internal" restraint, such as cable or column supports at intermediate points along a beam or plate.

We have already seen that redundant loads, most conveniently expressed as fractions of applied loads, can define independent variables of design. One heretofore obscure advantage of treating redundant loads as design variables is that the optimum values of redundant loads may be nonzero at the unloaded state. In such cases the optimum structure will require a self-equilibrating set of local internal loads even before the system environmental loads are applied. Such structures are referred to as *prestressed* and will be considered in Secs. 5 and 6.

As a first example of dissimilar-element redundancy consider the fundamental problem of column supports for an otherwise determinate beam system. In Chap. 5 we optimized such a combination by assuming provisionally a large enough number of equally spaced columns so that the two end supports could be neglected, thus yielding a determinate structure. We found there that for a given material the optimum column spacing was 17 percent of the column depth, indicating that if the beam span at least exceeded the column depth ($L > H$), there would at least be four column supports. Thus the neglecting of end supports would seem to be increasingly more accurate as L continued to exceed H. If $L < H$, however, the approach of Chap. 5 is unrealistic, and yet column supports may still represent improved efficiency. To explore this potential we must introduce redundant variables which define the load splits between end supports and columns. Such a redundant system employing a uniform simply supported beam and a single support column is shown in Fig. 6-8.

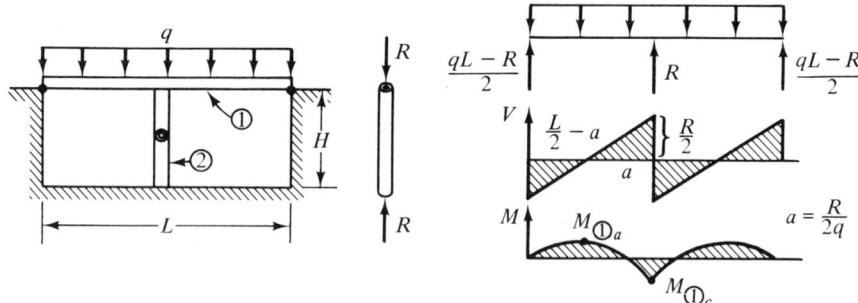

Fig. 6-8. Redundant Beam Column System

From the above shear and moment diagrams it can be seen that either of the two competing moments, M_a and M_c, may become critical. Introducing a redundant variable, k, such that the column load $R = kqL$, the magnitudes of these moments can be expressed as

$$M_a = (1-k)^2 \frac{qL^2}{8}$$

$$M_c = [k^2 - (1-k)^2] \frac{qL^2}{8}$$

(6-14)

The condition on k for which $M_a > M_c$ is thus

$$(1-k)^2 > k^2 - (1-k)^2$$

and yields

$$k < 0.586 \tag{6-15}$$

Specifying a uniform H-section beam and circular tube column, from Eqs. (2-44) and (4-37) the total system weight can be expressed as

$$W_s = \rho \frac{1.606}{E^{1/6}\sigma_{PL}^{1/2}} M_{\text{\textcircled{1}}}^{2/3} L + \rho \frac{H^3 \left(\frac{R}{H^2}\right)^{2/3}}{0.54 E^{2/3}} \tag{6-16}$$

where to simplify the evaluation we will consider only the case for which $\sigma_{A_{\text{opt}}} < \sigma_y$ and $\eta_T = 1$ for the column support. Substituting Eq. (6-14) for the beam local moment, with $R = kqL$, yields

$$W_s = \rho \frac{1.606}{E^{1/6}\sigma_{PL}^{1/2}} \left(\frac{qL^2}{8}\right)^{2/3} \{f(k)\}^{2/3} L + \rho \frac{H^3 \left(\frac{kqL}{H^2}\right)^{2/3}}{0.54 E^{2/3}} \tag{6-17}$$

where

$$f(k) = \begin{cases} (1-k)^2 & \text{for } k \leq 0.586 \\ k^2 - (1-k)^2 & \text{for } k \geq 0.586 \end{cases}$$

Since this form is similar to that of the determinate beam-column system of Chap. 5 a W_s/LH^2 index form is indicated and results in

$$\frac{W_s}{LH^2} = \rho \underbrace{\left[\frac{1.606}{(8)^{2/3} E^{1/6} \sigma_{PL}^{1/2}} \frac{\{f(k)\}^{2/3}}{(H/L)^{4/3}} + \frac{k^{2/3}(H/L)^{1/3}}{0.54 E^{2/3}}\right]}_{f_e} \left(\frac{q}{H}\right)^{2/3} \tag{6-18}$$

Since (q/H) factors, we are now in a position to optimize k independent of load environment with dependence on H/L only. Figure 6-9 shows the efficiency factor $f_e(k, H/L)$ plotted as a function of H/L for optimized k. Using AISI 1025 steel the optimum value of k is found to be 0.586 for $H/L < 2.9$ and zero for $H/L > 2.9$, yielding the conclusion that the redundant beam-column represents improved efficiency over the beam alone only for $H/L < 2.9$ and that in this region the redundant force split is always such that the competing moments, M_a and M_c, are equal.

The equivalent of $f_e(k, H/L)$ for the determinate beam-column system of Chap. 5 is also shown and indicates a significant weight savings for $H/L < 1$. The intersection of the two relationships just beyond $H/L = 1$ is seen to verify that the neglecting of end effects for the determinate system is unrealistic for $H/L > 1$.

Consider now a dissimilar-element system in which an orientation variable in addition to a redundant load is to be optimized. A representative problem is found in the cable-supported cantilever as shown in Fig. 6-10 where a uniform load distribution is to be carried along a line transverse to an existent wall.

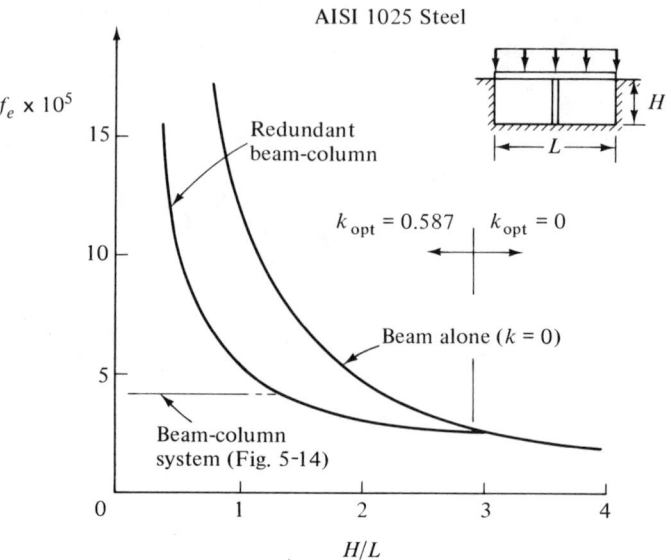

Fig. 6-9. Redundant Beam–Column Efficiency Evaluation in Terms of H/L

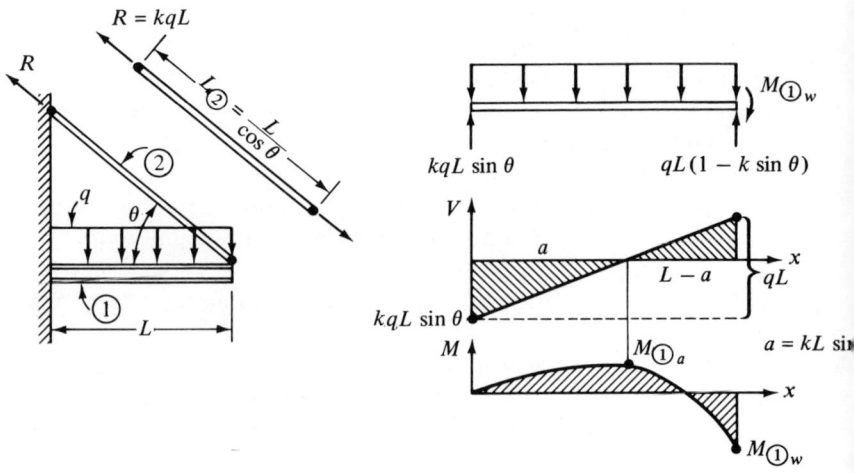

Fig. 6-10. Cable-Supported Cantilever Subjected to a Uniform Load Distribution

As in previous examples the redundant load carried by the cable R is expressed as kqL. Accordingly, from the shear and moment diagrams, the magnitudes of the two competing moments, M_a and M_w, can be expressed as

$$M_a = \frac{1}{2}k^2qL^2 \sin^2 \theta$$
$$M_w = \frac{1}{2}qL^2(1 - k \sin \theta)^2 - \frac{1}{2}k^2qL^2 \sin^2 \theta \tag{6-19}$$

Now determining k such that $M_a > M_w$ yields

$$k > \frac{1}{(1 + \sqrt{2}) \sin \theta} \tag{6-20}$$

The total system weight for a uniform H-section beam and axial cable can be written as

$$W_s = \rho \frac{1.606}{E^{1/6}\sigma_{PL}^{1/2}} M_①^{2/3} L + \rho \frac{RL_②}{\sigma_y} \tag{6-21}$$

where from Fig. 6-10, $L_② = L/\cos \theta$, $R = kqL$, and where $M_①$ is the larger of M_a and M_w in Eq. (6-19). Making these substitutions yields

$$W_s = \rho \frac{1.606}{E^{1/6}\sigma_{PL}^{1/2}} (qL^2)^{2/3} \{f(k, \theta)\}^{2/3} L + \rho \frac{kqL\left(\dfrac{L}{\cos \theta}\right)}{\sigma_y} \tag{6-22}$$

where

$$f(k, \theta) = \begin{cases} \dfrac{1}{2}k^2 \sin^2 \theta & \text{for } k \geq \dfrac{1}{(1 + \sqrt{2}) \sin \theta} \\ \dfrac{1}{2}(1 - k \sin \theta)^2 - \dfrac{1}{2}k^2 \sin^2 \theta & \text{for } k \leq \dfrac{1}{(1 + \sqrt{2}) \sin \theta} \end{cases}$$

Dividing through by L^3, Eq. (6-22) reduces to index form

$$\frac{W_s}{L^3} = \rho \left[\frac{1.606}{E^{1/6}\sigma_{PL}^{1/2}} \{f(k, \theta)\}^{2/3} \left(\frac{q}{L}\right)^{2/3} + \frac{k}{\sigma_y \cos \theta}\left(\frac{q}{L}\right)\right] \tag{6-23}$$

A numerical evaluation of Eq. (6-23) for AISI 1025 steel results in the same optimum pair of θ and k (49° and 0.55 respectively) for all q/L less than 680 psi.* Fig. 6-11 shows a plot of weight index in terms of q/L which compares Eq. (6-23) at optimized θ and k to the cantilever alone ($k = 0$). Note that the optimum values of k and θ correspond to equal magnitudes of competing moments, viz., $0.55 = 1/[(1 + \sqrt{2}) \sin 49°]$. It is also found that if θ is changed moderately from its optimum value, the weight penalty is practically nil as long as a k is selected such that $k = 1/[(1 + \sqrt{2}) \sin \theta]$

---

*At large q/L the cable weight dominates and eventually the beam alone ($k = 0$) is optimum.

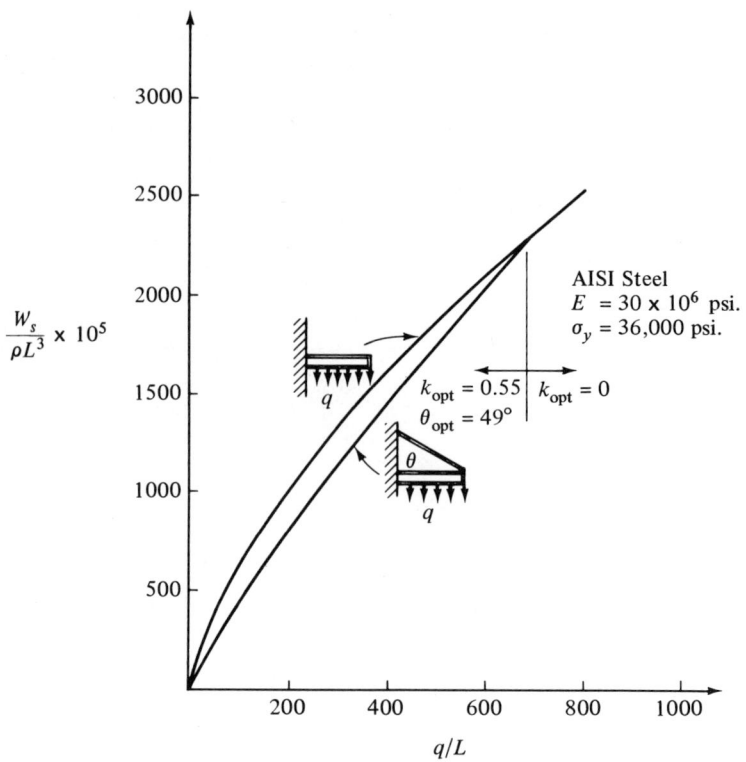

Fig. 6-11. Weight Index Versus Load Index for Redundant Cantilever Cable System

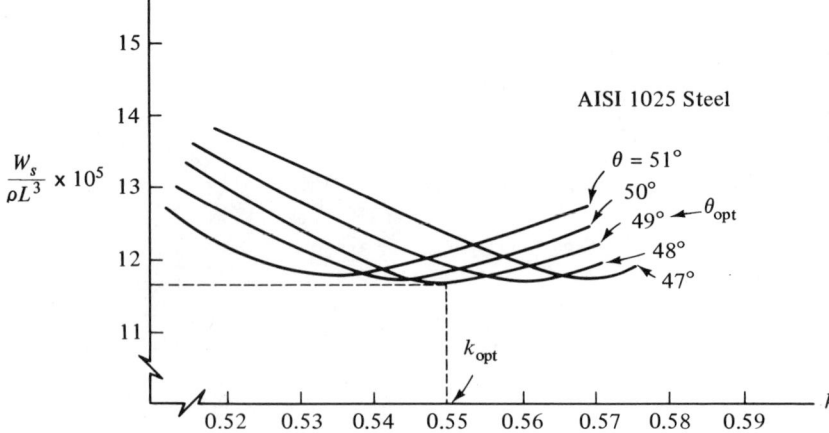

Fig. 6-12. Weight Sensitivity in Terms of θ and k at $q/L = 1$ for Redundant Cantilever–Cable System

(the condition for equal competing moments). This is illustrated in Fig. 6-12 where for q/L fixed at unity, the weight index is plotted in terms of θ and k.

The sensitivity analysis of Fig. 6-12 tells us that total structural weight is practically independent of θ as long as a corresponding value of k is selected so that competing moments are equalized. Recall at this point that in the examples considered where either a beam, column, or cable provided an optimized redundant support to an otherwise determinate beam, optimum conditions dictated equal competing moments.

Section 4
COMPATIBILITY

The loads *analysis* of a given redundant structure requires either compatibility of deformation or an energy principle equivalent in addition to the equations of equilibrium. In the *synthesis* of redundant structures we have not employed any such supplementation to equilibrium. Rather we have required the redundant load variables to take on values which maximize merit with all remaining "determinate" loads following from equilibrium requirements. As such the design solution has the evident properties that: (1) redundant supports or elements carry an optimized state of load, and (2) equilibrium is satisfied. Certainly from the design viewpoint such a solution as an ideal represents an optimum within the candidates of the specified form.

There remains the practical question of whether such a structure, once realized and sustaining the applied loads, will be found to yield a deformed state where redundant supports or elements will in fact carry their optimum loads. Since compatibility is in no way employed in the synthesis of configuration, is it reasonable to expect that the structure, as a result of its inherent compatible deformations, will assume the optimized state of local loads?

The answer to this question is to be found in the most fundamental property of redundant structures. This property is that redundancy always represents an *excess* of support or elements over an otherwise static configuration. As such we can always adjust either the flexibility of redundant supports or the lengths of redundant elements so they will carry that portion of the applied loads which corresponds to an optimized state. For example, in the cable-cantilever of Fig. 6-11 it was found that the redundant cable must carry 55 percent of the applied load magnitude, qL, for a minimum in total weight of the combination. When such a configuration is fabricated, any representative uniformly distributed load could be applied to the beam (the weight of the beam itself could suffice) and the cable support, using a turnbuckle or other device, could be adjusted until it carried 55 percent of the representative load. Due to the linear response of redundant load to applied load (small deflection analysis) it would therefore be found at the applied load that the local loads for beam and cable would be at their optimized values.

Hence, by physical adjustment of the initial condition (length or preset load) in a redundant element, the optimized state of load for the total structure can be obtained. By the fundamental property of redundancy such adjustments can always be effected on the realized structure and, therefore, in addition to guarantees of equilibrium and maximized merit, compatibility of deformation is assured.

Section 5
REDUNDANT PRESTRESSED SYSTEMS

The use of redundant prestressed elements which carry appreciable loads at the no-load condition represents an important avenue of investigation in the search for optimum form. Aside from minor interfacing internal loads (such as discussed in Sec. 4), the redundant elements of previous examples were essentially unstressed at the no-load condition.

In the design of concrete beam structures, prestressed steel bars have often been employed in an effort to eliminate tensile stress and increase the load-carrying capability of the composite. The use of prestressed redundancy in such instances, however, has primarily been the result of concrete's inability to carry appreciable tensile stress. The potential for improved efficiency using prestressing of metal structures, where both tension and compression can be equally sustained, has been relatively unexplored in conventional structures.

Actually, there can be potential weight savings in such metal structures using the prestressed concept. To illustrate this, consider the example of Fig.

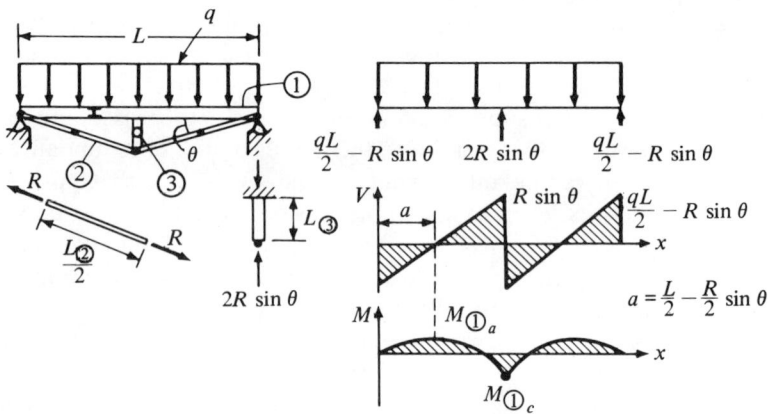

Fig. 6-13. Beam and Prestressed Cable Redundant System

6-13, where a beam and prestressed cable-column system is to be optimized for minimum weight subject to a uniform load distribution.

The system utilizes a uniform H-section beam, cable, and uniform circular tube column with fixed-free end supports in a redundant prestressed arrangement as shown. R represents an axial preset load carried by the cable which prestresses the entire structure at the no-load condition. The free bodies shown in the figure represent the respective local load states at the applied loads and hence, strictly speaking, the cable prestress must be established by adjustment at this condition. We also introduce an orientation variable θ as shown in the figure.

Representing R as kqL,* the local load for the beam can be expressed in terms of two competing moment magnitudes as

$$M_a = \frac{qL^2}{8}(1 - 2k \sin \theta)^2$$
$$M_c = \frac{qL^2}{2}k^2 \sin^2 \theta - \frac{qL^2}{8}(1 - 2k \sin \theta)^2 \tag{6-24}$$

where for $M_a > M_c$ it is found that

$$k < \frac{1}{(\sqrt{2} + 2) \sin \theta} \tag{6-25}$$

Cable and column local loads can be expressed in terms of k as follows

$$P_② = R = kqL \tag{6-26}$$
$$P_③ = 2kqL \sin \theta \tag{6-27}$$

From the figure the local lengths can be evaluated as

$$L_① = L \tag{6-28}$$
$$L_② = \frac{L}{\cos \theta} \tag{6-29}$$
$$L_③ = \frac{L}{2} \tan \theta \tag{6-30}$$

The total weight for a beam, cable, column combination can be written as

$$W_s = \rho \frac{1.606}{E^{1/6}\sigma_{PL}^{1/2}} M_①^{2/3} L_① + \rho \frac{P_② L_②}{\sigma_y} + \rho \frac{L_③^3 (P/L_③^2)^{2/3} c^{2/3}}{0.54 E^{2/3}} \tag{6-31}$$

where to simplify the evaluation we will consider only the case for which $\sigma_{A_{opt}} < \sigma_y$ and $\eta_T = 1$ for the column post. Also we neglect the axial thrust carried by the beam due to the cable loads. This load could be accounted for in a *post factum* modification similar to the shear web modification required for a beam optimized to carry pure bending.

*Although we represent R as a percentage of qL for mathematical convenience, recognize that the cable load exists as a nonzero preset load even when q is not applied.

Introducing the local loads and lengths, Eqs. (6-24) through (6-30) into Eq. (6-31) yields

$$W_s = \rho \frac{1.606}{E^{1/6}\sigma_{PL}^{1/2}}(qL^2)^{2/3}\{f(k,\theta)\}^{2/3}L + \rho\frac{kqL^2}{\sigma_y \cos\theta}$$
$$+ \rho\frac{\left(\frac{L}{2}\tan\theta\right)^{5/3}(2kqL\sin\theta)^{2/3}(2)^{2/3}}{0.54E^{2/3}}$$

where

$$f(k,\theta) = \begin{cases} \frac{1}{8}(1-2k\sin\theta)^2 & \text{for } k \leq 1/[(2+\sqrt{2})\sin\theta] \\ \frac{1}{2}k^2\sin^2\theta - \frac{1}{8}(1-2k\sin\theta)^2 & \text{for } k \geq 1/[(2+\sqrt{2})\sin\theta] \end{cases} \quad (6\text{-}32)$$

and dividing through by L^3 results in

$$\frac{W_s}{L^3} = \rho\left[\frac{1.606}{E^{1/6}\sigma_{PL}^{1/2}}\{f(k,\theta)\}^{2/3}\left(\frac{q}{L}\right)^{2/3} + \frac{k}{\sigma_y\cos\theta}\left(\frac{q}{L}\right) + \frac{k^{2/3}\sin^{2/3}\theta\tan^{5/3}\theta}{0.54E^{2/3}(2)^{1/3}}\left(\frac{q}{L}\right)^{2/3}\right] \quad (6\text{-}33)$$

A numerical evaluation of Eq. (6-33) for AISI 1025 steel yields the same optimum pair of k_{opt} and θ_{opt} (0.51 and 35° respectively) when the load index is less than 95 psi. Again these values correspond to equality of the competing moments in the beam at the applied and preset loads. Fig. 6-14 (p. 119) shows a comparison between the prestressed beam and the beam alone.

In the example above, if the single applied load condition is not dead weight, the optimum orientation and redundancy must be evaluated by also considering the internal preload as a *separate* load condition. This aspect of redundant systems design will be illustrated in the following section.

Section 6

GENERAL APPROACH FOR TRUSSES UNDER MULTIPLE LOAD CONDITIONS*

Let $|P_{i,j}|$ represent the magnitude of the internal bar load for the ith bar under the jth external load condition and $\sigma_{opt_{i,j}}$, the magnitude of the maximum stress that can be developed in the ith bar under the jth load condition. Obviously, considering all loading conditions, the area of the ith bar must be selected as $(|P_{ij}|\sigma_{opt_{i,j}})_{max}$. Therefore the minimum truss weight can be expressed by

*Part of a paper published in the *Proc. ASCE* 96, ST 12, Dec., 1970.

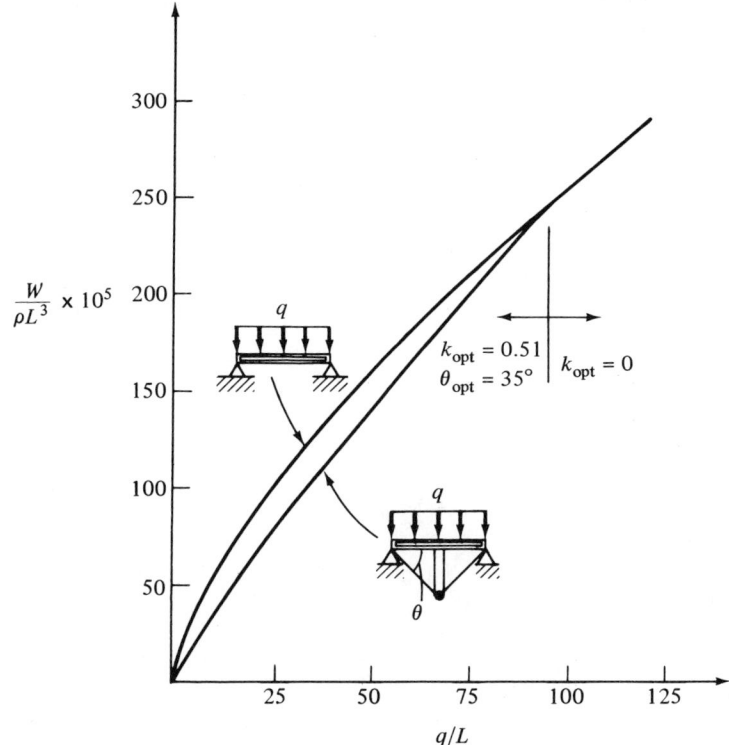

Fig. 6-14. Weight Index Versus Load Index for Prestressed Beam System

$$W_s = \sum_i \rho \left(\frac{|P_{i,j}|}{\sigma_{\text{opt}_{i,j}}} \right)_{\max} L_i \tag{6-34}$$

where for tensile loads $\sigma_{\text{opt}_{i,j}} = \sigma_y$ (yield stress) and for compressive loads

$$\sigma_{\text{opt}_{i,j}} = \text{the least of} \begin{cases} 0.54\, \eta_T^{1/2} E^{2/3} \left(\dfrac{|P_{i,j}|}{L_i^2} \right)^{1/3} \\ \sigma_y \end{cases} \tag{6-35}$$

assuming circular tube columns (a similar expression would result for other column cross sections). To show that Eq. (6-34) can be minimized with respect to orientation variables and redundant loads within an index formulation, proceed as follows. Express each of the externally applied and reactive loads (under all loading conditions) as a fraction of any one of the given loads. Define this load as P and, in addition, express each redundant load as a variable fraction of P, i.e., $R_1 = k_1 P$, $R_2 = k_2 P$, etc. It follows from the equations of equilibrium that for the jth load condition

$$|P_{i,j}| = P|p_{i,j}| \tag{6-36}$$

where $p_{i,j}$ is a function of redundant load variables, k, and orientation variables, S_o, only. The function $p_{i,j}$ which represents the extreme condition is that for which $|p_{i,j}|/\sigma_{\text{opt}_{i,j}}$ is maximum. This guarantees a nonfailing cross section under all load conditions since

$$A_i = (P|p_{i,j}|/\sigma_{\text{opt}_{i,j}})$$

and P is fixed.

Further, express each of the fixed geometric lengths given in the design problem as a fraction of any one of these lengths. Defining this length as L, it follows that each of the component bar lengths, L_i, can be written as

$$L_i = L l_i \tag{6-37}$$

where l_i is a function of orientation variables only.

Combining Eqs. (6-34), (6-36), and (6-37) and dividing through by L^3 yields an index form for the weight merit.

$$\frac{W_s}{L^3} = \left[\sum_i \rho \left(\frac{|p_{i,j}|}{\sigma_{\text{opt}_{i,j}}} \right)_{\max} l_i \right] \frac{P}{L^2} \tag{6-38}$$

The column-yield test condition, Eq. (6-35), can also be put in index form by combining with Eqs. (6-36) and (6-37), and yields

$$\sigma_{\text{opt}_{i,j}} = \text{the least of} \begin{cases} 0.54 \, \eta_T^{1/2} E^{2/3} \left(\frac{|p_{i,j}|}{l_i^2} \right)^{1/3} \left(\frac{P}{L^2} \right)^{1/3} \\ \sigma_y \end{cases} \tag{6-39}$$

Note since each $p_{i,j}$ and l_i is a function of S_o and k, Eqs. (6-38) and (6-39) represent a general index formulation where the only independent design variables are the S_o's and k's.

Application to a Redundant Truss under a Single Load Condition

Consider the conventional Howe roof truss under three symmetric equal loads as shown in Fig. 6-15. The basic determinate truss is shown in Fig. 6-15a where orientation is completely defined in terms of the angle α. In Fig. 6-15b a redundant modification of the Howe truss is shown which introduces the additional variables θ and R. Only the weight merit for the redundant truss will be formulated since the determinate configuration is a special case at $R = 0$.

Note as shown in the figure that each of the applied and reactive loads are expressed as a fixed fraction of P and that the redundant load has been expressed as a variable fraction of P. From geometry the length of each component bar can be expressed as a product of the fixed span, L, and a function of orientation variables, α and θ, as follows

$$L_{1-2} = \left(\frac{1}{4\cos\alpha}\right)L = L_{2-3} = L_{3-4} = L_{4-5} = L_{2-7} = L_{7-4}$$

$$L_{1-8} = \left(\frac{1}{4}\right)L = L_{8-7} = L_{7-6} = L_{6-5}$$

$$L_{2-8} = \left(\frac{\tan\alpha}{4}\right)L = L_{4-6}$$

$$L_{3-7} = \left(\frac{\tan\alpha}{2}\right)L \qquad (6\text{-}40)$$

$$L_{1-9} = \left(\frac{1}{2\cos\theta}\right)L = L_{5-9}$$

$$L_{7-9} = \left(\frac{\tan\theta}{2}\right)L$$

where L_{1-2} denotes the length of the member between nodes 1 and 2, etc.

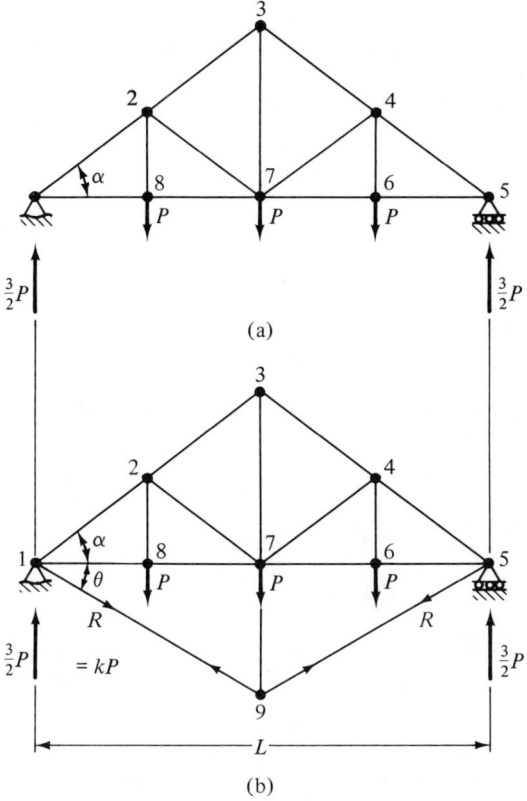

Fig. 6-15. Determinate and Redundant Howe Truss under a Single Load Condition

Although strictly speaking there is a single load condition, recognize that a given value of the redundant load will—through possible prestress— create an internal load condition for all bars in the absence of the externally applied load. For some bars (depending on the design point $\{\alpha, \theta, k\}$) this internal prestressed load condition may represent the extreme. Hence the required area for all bars must be evaluated by considering both the applied load (in conjunction with the redundant load) and the redundant load alone as separate load conditions (certainly this is not essential if the single external load condition is dead weight).

Defining the combination of the external load condition and the internal redundant load as the 1 condition, we find (from equilibrium of concurrent forces at nodes 1, 2, 3, and 9) each component bar load as a product of P and a function of α, θ, and k as follows

$$P_{1-2,1} = \left[\frac{(k \sin \theta - 3/2)}{\sin \alpha}\right] P = P_{4-5,1}$$

$$P_{1-8,1} = \left[- k \cos \theta - \frac{(k \sin \theta - 3/2)}{\tan \alpha}\right] P = P_{6-5,1} = P_{8-7,1} = P_{7-6,1}$$

$$P_{2-3,1} = \left[\frac{(k \sin \theta - 1)}{\sin \alpha}\right] P = P_{3-4,1}$$

$$P_{2-7,1} = \left[- \frac{1}{2 \sin \alpha}\right] P = P_{7-4,1} \tag{6-41}$$

$$P_{2-8,1} = P = P_{4-6,1}$$

$$P_{3-7,1} = [- 2(k \sin \theta - 1)] P$$

$$P_{1-9,1} = kP = P_{5-9,1}$$

$$P_{7-9,1} = [- 2k \sin \theta] P$$

where $P_{1-2,1}$ denotes the axial bar load in the member between nodes 1 and 2 under load condition 1, etc.

For the internal redundant load condition (condition 2) we find

$$P_{1-2,2} = \left[\frac{k \sin \theta}{\sin \alpha}\right] P = P_{4-5,2}$$

$$P_{1-8,2} = \left[- k \cos \theta - \frac{k \sin \theta}{\tan \alpha}\right] P = P_{6-5,2} = P_{8-7,2} = P_{7-6,2}$$

$$P_{2-3,2} = \left[\frac{k \sin \theta}{\sin \alpha}\right] P = P_{3-4,2}$$

$$P_{2-7,2} = 0 = P_{7-4,2} \tag{6-42}$$

$$P_{3-7,2} = [- 2k \sin \theta] P$$

$$P_{1-9,2} = kP = P_{5-9,2}$$

$$P_{7-9,2} = [- 2k \sin \theta] P$$

Note that for $k = 0$ each bar load at condition 2 is zero.

The total truss weight is now expressed as

$$W_s = \sum_i \rho \left(\frac{|P_{i,j}|}{\sigma_{\text{opt}_{i,j}}}\right)_{\max} L_i \qquad (6\text{-}43)$$

By Eqs. (6-40), (6-41), and (6-42) each $P_{i,j}$ and L_i can be expressed as $P_{i,j} = |p_{i,j}| P$ and $L_i = l_i L$ where $p_{i,j}$ and l_i are functions of α, θ, and k only. Therefore Eq. (6-43) can be rewritten as

$$W_s = \left[\sum_i \rho \left(\frac{|p_{i,j}|}{\sigma_{\text{opt}_{i,j}}}\right)_{\max} l_i\right] PL \qquad (6\text{-}44)$$

and dividing by L^3 yields

$$\frac{W_s}{L^3} = \left[\sum_i \rho \left(\frac{|p_{i,j}|}{\sigma_{\text{opt}_{i,j}}}\right)_{\max} l_i\right] \frac{P}{L^2} \qquad (6\text{-}45)$$

(which matches the general formulation of Eq. (6-38)). Eq. (6-45) is in index form since $\sigma_{\text{opt}_{i,j}}$ is constant at yield for tensile loads ($p_{i,j} > 0$) and a function of P/L^2 for compressive loads ($p_{i,j} < 0$). (See Eq. (6-39))

Figure 6-16 shows an index comparison of the redundant configuration and the determinate configuration (obtained by setting $k = 0$). The computer solution was accomplished by iterative search for the optimum point $\{\alpha, \theta, k\}_{\text{opt}}$ for a specified material and various load indices. Note that for any trial design point $|p_{i,j}|/\sigma_{\text{opt}_{i,j}}$ can be evaluated for each of the two load conditions and the larger selected. Other than this test condition the search is seen to be unconstrained in the classical sense since all practically realizable permutations of α, θ and k are acceptable design points. The comparison shown in Fig. 6-16 is for AISI 1025 steel, $E = 30 \times 10^6$ psi, $\sigma_y = 36{,}000$ psi, and $\rho = 0.283$. It can be noted that the redundant prestressed truss represents a fixed percentage weight savings (13.3 percent) over the determinate configuration for all load indices $P/L^2 > .05$. Above $P/L^2 = .05$ all compression members achieve yield stress and hence the weight curves are linear. Below this index compression members are buckling designed but the 13 percent weight difference is roughly maintained. Also shown in a comparison with a determinate circular tube beam (obtained in index form by the method of the previous chapter as $W/L^3 = 3.97\rho(P/L^2)^{2/3}/(E\sigma_y)^{1/3}$). Note since the beam weight grows as $(P/L^2)^{2/3}$ an index must eventually be reached where the beam is lighter than either the determinate or redundant truss configurations.

An example evaluation for the specific values $P = 10{,}000$ lb, $L = 100$ lb is shown in Figs. 6-17 and 6-18. An analysis of the determinate and redundant trusses shown in these figures verifies that each bar satisfied the buckling and yield constraints and that the weight difference is as given in the index comparison of Fig. 6-16. This analysis also shows that all bars are fully stressed under the combination of the external and redundant load condition but only bars 1–8, 8–7, 7–6, 6–5, 1–9, 7–9, and 5–9 are fully stressed under the internal prestress load condition alone.

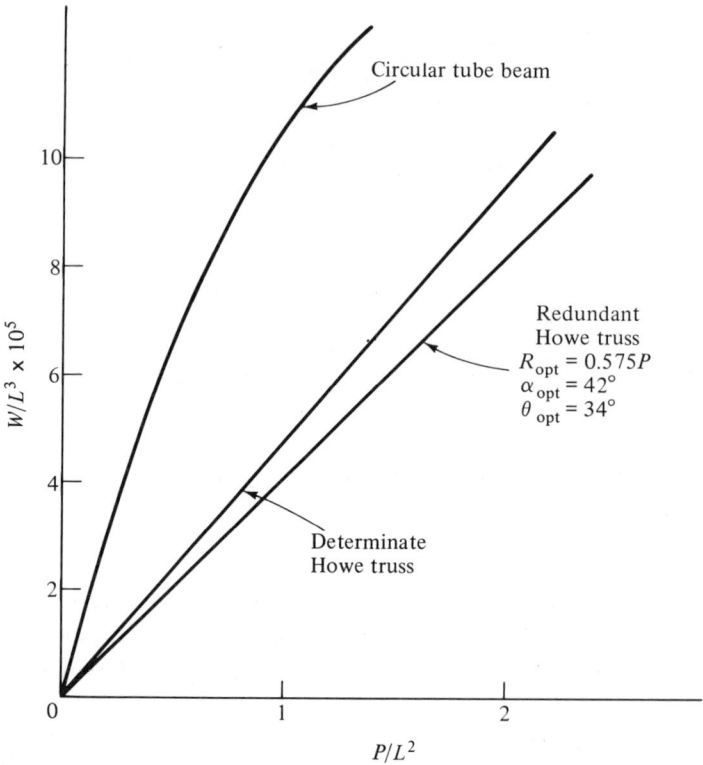

Fig. 6-16. Index Comparison of a Determinate and Redundant Howe Truss for AISI 1025 Steel

Sved[17] showed that the minimum weight truss is determinate under a single load condition. Although Sved allows for prestressed redundant loads he does not consider the internal preload as a separate load condition. The present analysis verifies that if one were to exclude the internal prestressing as a separate load condition a determinate truss would result at minimum weight. This determinate solution, however, would become a mechanism under the action of the internal prestressing alone, since the horizontal members 1–8, 8–7, 7–6, and 6–5 vanish. Without these members, the truss will obviously collapse in the absence of the external load condition.

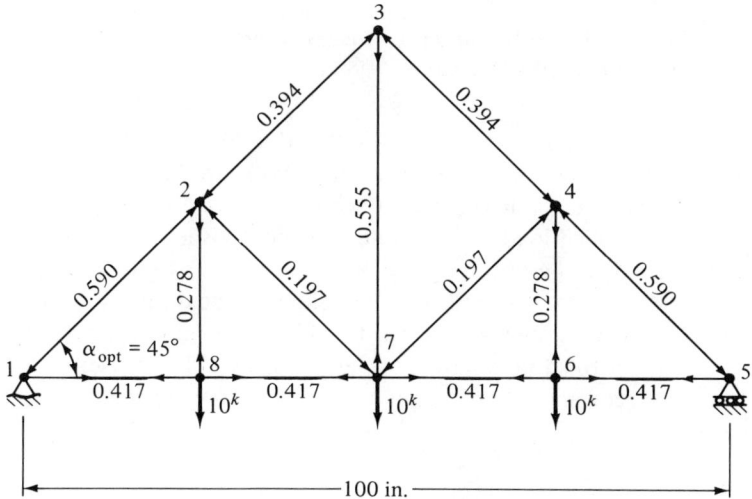

Fig. 6-17. Optimized Determinate Howe Truss for $P = 10^K$, $L = 100$ in and AISI 1025 Steel ($W_s = 47.2$ lb)

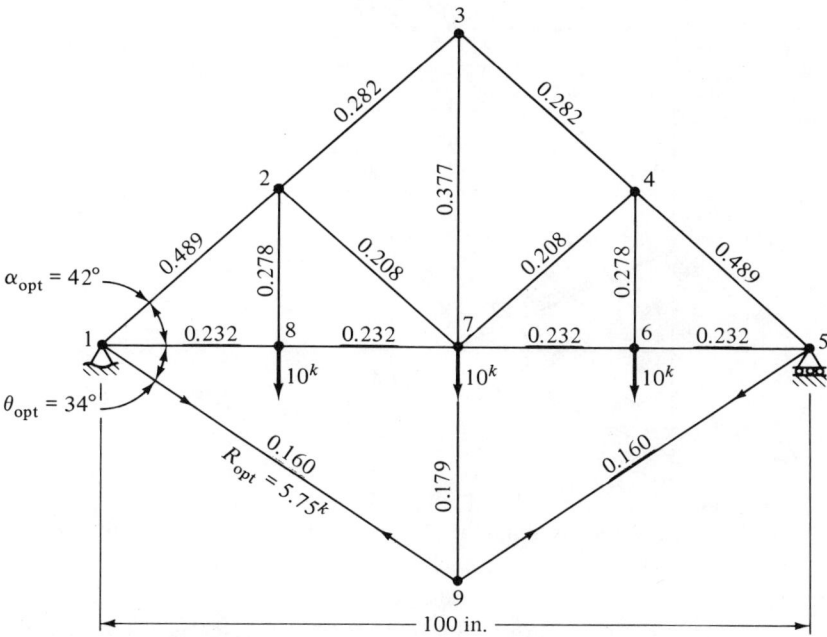

Fig. 6-18. Optimized Redundant Howe Truss for $P = 10^K$, $L = 100$ in and AISI 1025 Steel ($W_s = 40.7$ lb)

Application to Redundant Trusses under Multiple Load Conditions

Consider the trusses shown in Figs. 6-19 and 6-20. In Fig. 6-19 the displacement of node 2 must be symmetric under both applied loads. As a consequence a single optimum value of the prestress load can be maintained for either of the applied loads. This could be accomplished by employing a turnbuckle adjusted to the optimum prestress under the action of any one of the loads. For the truss shown in Fig. 6-20 a single value of the optimum prestress could not be maintained with a turnbuckle since node 2 displaces unsymmetrically under the different loads. Here a turnbuckle in series with a flexible spring could reasonably approximate the same prestress with differing displacements.

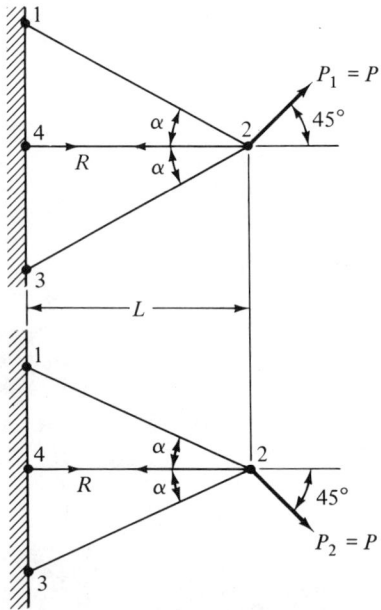

Fig. 6-19. Redundant Truss under Symmetrical Multiple Loads

The load and length evaluations for these cases as a function of orientation variables and redundant load are obtained in like fashion to the previous application and will not be listed. Note, however, that here there are actually three load conditions since the redundant load must be considered as a separate load condition (this is essential since the structure sustains no external load in the transition from P_1 to P_2).

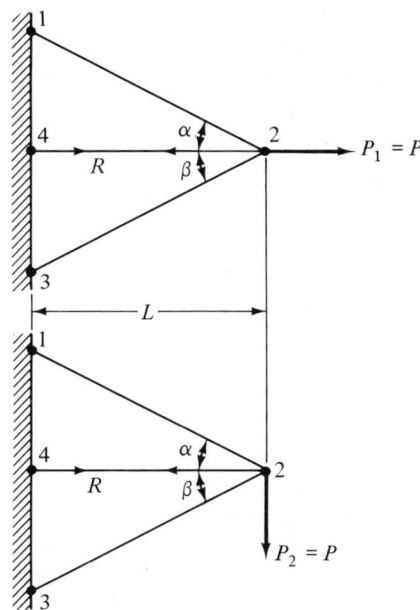

Fig. 6-20. Redundant Truss under Unsymmetrical Multiple Loads

Results of the optimization of these examples are shown in Figs. 6-21, 6-22, 6-23, and 6-24. The figures show the optimum values of the independent variables (orientation and redundancy) and dependent variables (cross-sectional areas of each bar) for the fixed load and geometry shown and AISI 1025 steel ($\sigma_y = 36{,}000$ psi). The percentage weight savings (redundant compared to the determinate) are 16.2 and 7.03 for the symmetric and unsymmetric load cases respectively.

For the symmetric case it was found that all bars were fully stressed under either external load condition and understressed for the internal redundant load condition alone. Also, compression members were buckling designed under both load conditions (member 1-2 under P_1 and member 2-3 under P_2).

Schmit[4] presents the same example with fixed orientation ($\alpha = 45°$) and redundant load following from compatibility and finds that only a single bar (that collinear with the current load) is fully stressed at each load condition achieving a 6.7 percent weight savings over the determinate case.

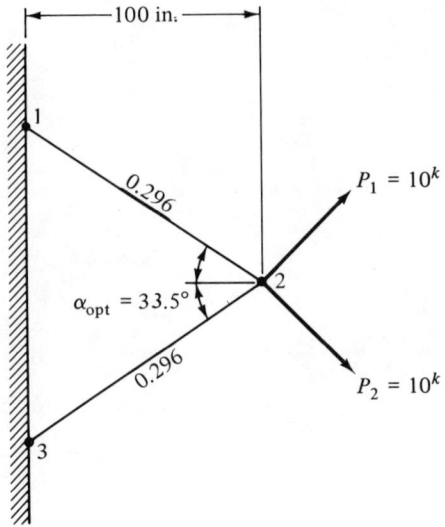

Fig. 6-21. Optimized Determinate Truss under Symmetrical Multiple Loads for $P_1 = P_2 = 10^k$, $L = 100$ in and AISI 1025 Steel ($W_s = 20.1$ lb)

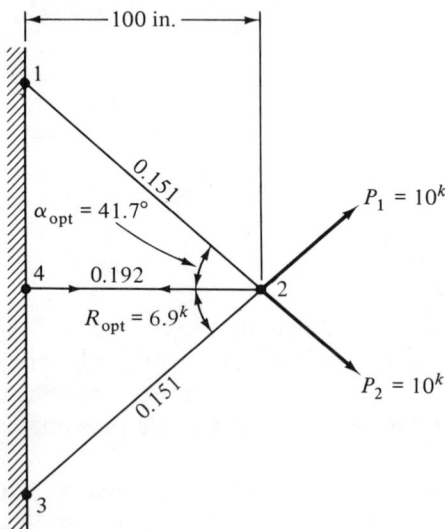

Fig. 6-22. Optimized Redundant Truss under Symmetrical Loads for $P_1 = P_2 = 10^k$, $L = 100$ in and AISI 1025 Steel ($W_s = 16.8$ lb)

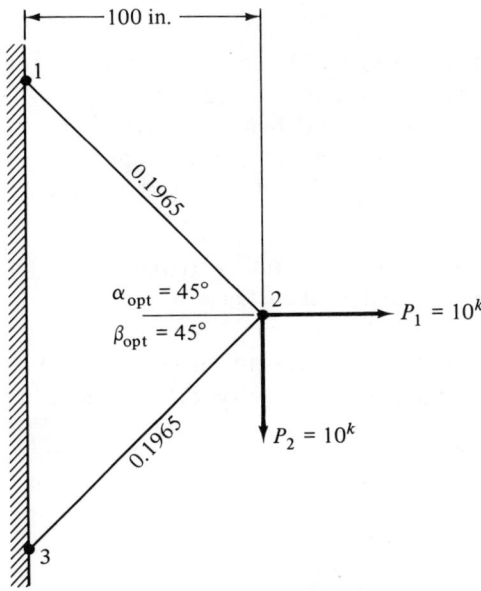

Fig. 6-23. Optimized Determinate Truss under Unsymmetrical Multiple Loads for $P_1 = P_2 = 10^K$, $L = 100$ in and AISI 1025 Steel ($W_s = 16.9$ lb)

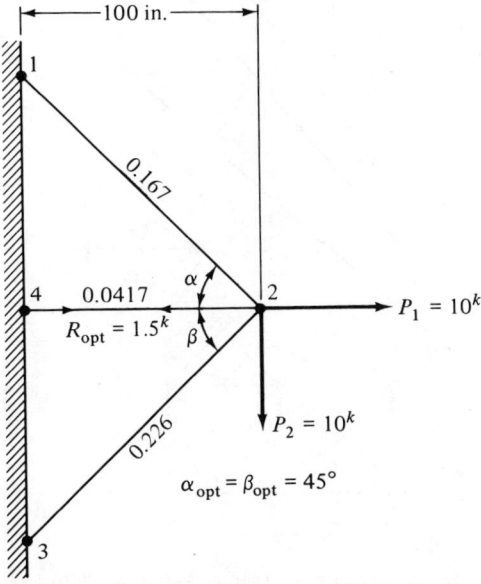

Fig. 6-24. Optimized Redundant Truss under Unsymmetrical Multiple Load for $P_1 = P_2 = 10^K$, $L = 100$ in and AISI 1025 Steel ($W_s = 15.7$ lb)

Therefore the present example shows an additional 9.5 percent weight savings by optimizing bar orientation and redundant load. For the unsymmetric case all bars are fully stressed for P_2 but bar 2–3 is understressed for P_1 (again all bars are understressed for the internal redundant load condition). The single compression member (bar 2–3 under P_2) was yield designed.

Index Comparison of a Prestressed Conventional Truss with the Lower Bound Michell Truss

Cox[18] shows that the following equation gives the lower bound Michell truss weight for a single applied load with two symmetric reactions all at a common elevation.

$$W_s = \rho \frac{1.285\,PL}{\sigma_y} \tag{6-46}$$

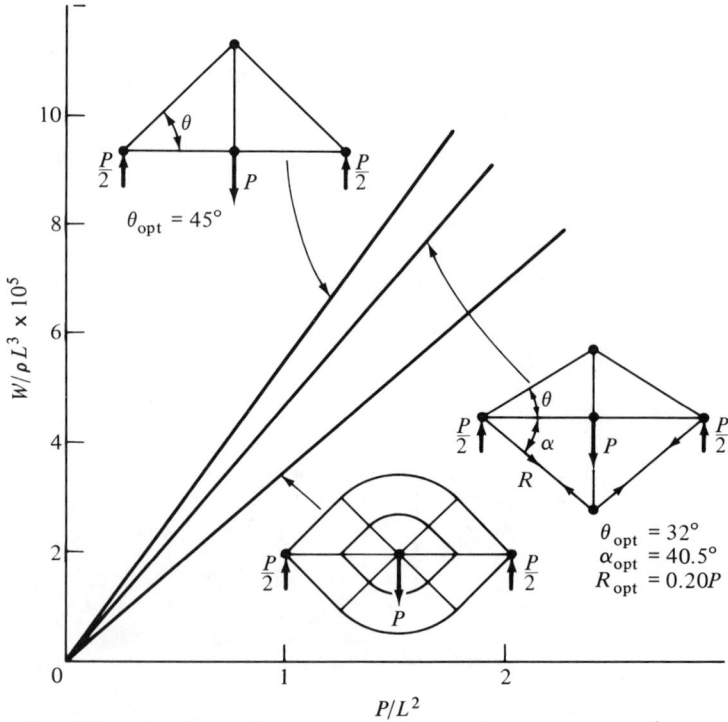

Fig. 6-25. Index Comparison of a Determinate and Prestressed Redundant King Pin Truss with the Michell Lower Bound Network under a Single Load Condition Using AISI 1025 Steel

An index form is generated by dividing through L^3 and yields

$$\frac{W_s}{L^3} = \rho \frac{1.285}{\sigma_y}\left(\frac{P}{L^2}\right) \tag{6-47}$$

Fig. 6-25 shows an index comparison of the Michell truss and the conventional King Pin truss with determinate and prestressed configurations (ignoring buckling constraints). Index expressions for the King Pin truss were obtained by the general truss approach previously described. This yields for the determinate and redundant cases respectively

$$\frac{W_s}{L^3} = \rho \frac{2.00}{\sigma_y}\left(\frac{P}{L^2}\right) \tag{6-48}$$

$$\frac{W_s}{L^3} = \rho \frac{1.73}{\sigma_y}\left(\frac{P}{L^2}\right) \tag{6-49}$$

The comparison of Fig. 6-25 shows that the Michell truss is 35.6 percent lighter than the determinate King Pin truss with the redundant modification giving a 13.5 percent weight savings over the determinate case.

PROBLEMS

6-1. A uniform beam of length L under a uniform load distribution is to be supported at two symmetric points along its length. Show that the beam will be of least weight (and under equal competing moments) when the distance between these supports is $0.586\,L$.

6-2. A uniform beam with a center span load, P, can be supported at both ends. Due to the nature of existant material geometry, only the left support can be designed to restrain against rotation. Evaluate the optimum redundant moment M_R as a function of PL. How does this compare to a fixed (zero slope) restraint?

ANSWER $\quad M_R = \dfrac{PL}{6}.$

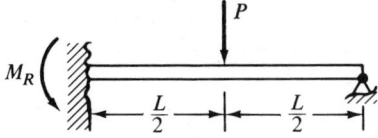

6-3. Determine the maximum value of H/L for which the redundant beam system shown is more efficient than a single beam. Assume pinned ends.

ANSWER $H/L = 0.75$.

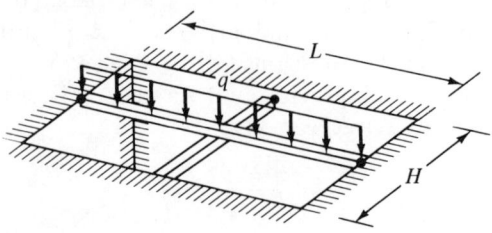

6-4. A uniform circular tube antenna is to be cantilevered from the fuselage of an airplane and supported by a redundant guy wire as shown. At maximum velocity the drag force results in q(lb/in) uniformly distributed over the tube. Determine the optimum prestressed guy wire load as a fraction of qL such that the tube will be of minimum weight.

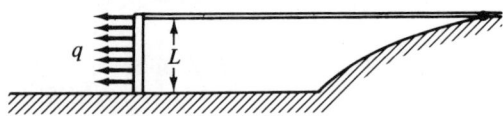

6-5. Show that the three-bar redundant truss of Fig. 5-2 reduces to a determinate form at optimum conditions. Make the evaluation for the fully stressed condition.

6-6. A beam column system as shown is to be optimized for minimum weight. Develop the relationships for determining the optimum values of column placement a and redundant load R. Can both these quantities be determined solely from a requirement of equal competing moments?

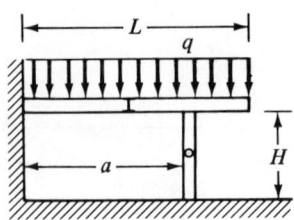

7
ADDITIONAL PROBLEMS IN CONVENTIONAL STRUCTURAL DESIGN

In this chapter we apply the methodology of design to some of the conventional load environments and structural forms found in both aeronautical and civil applications. Although we have already considered typical load environments, we have focused only on those forms which constitute a system combination of slender cables, columns and beams. There was a good reason for limiting our study of structural form to these basic elements; to plunge prematurely into more complex candidate forms would have run the risk of obscuring the methodological approach to design which was concurrently being established. However, once a general methodology has been established it then becomes appropriate to apply it to more sophisticated examples. Just as the simplified introductory applications served to build a methodology, the more advanced applications can often extend it.

The examples dealt with in this chapter, therefore, will serve the two objectives of extending design methodology while exposing the student to some of the contemporary structural design problems found in engineering.

Section 1

MINIMUM WEIGHT WIDE COLUMNS

The wide-column environment is a distributed compressive line load, q lb/in, transverse to an axial transmission path of length L (see Fig. 7-1). We define the general *wide-column form* as a two-dimensional panel configuration with a width equal to the line load distribution and a length equal to the axial transmission path. The panel may be stiffened with attached or integral stiffener elements, may be "folded" to produce a corrugation stiffening effect, may be of "sandwich" construction utilizing twin facesheets separated by a web or honeycomb core, etc. Examples of how a wide-column panel can arise in engineering applications can be seen in the compression surface of a rib-supported wing box, or a frame-supported large diameter fuselage shell (the shell curvature can usually be neglected for moderately large diameter shells). In these applications a distributed panel is necessary to maintain a continuous surface for airfoil requirements.

Assessment of Environment As in the case of the slender column, the wide-column panel exhibits the inherent failure mode of Euler buckling. Here, however, the buckling along the path of least resistance will always be about the axis parallel to the load distribution and it is the moment of inertia about this specific axis that must be made large for a given cross-sectional area. Hence the problem of wide-column form specification is similar to that of bending moment transmission utilizing a wide beam. We know that an efficient slender beam results with flanges and a web shear tie. Extrapolating this to a panel configuration would logically lead to a panel with flanged web stiffeners or a folded panel where the flats parallel to the buckling axis would serve as flanges with flats askew to this axis serving as webs.

Flat Corrugation Wide Column

As an example consider the three proportion variable flat corrugation cross section shown in Fig. 7-1.

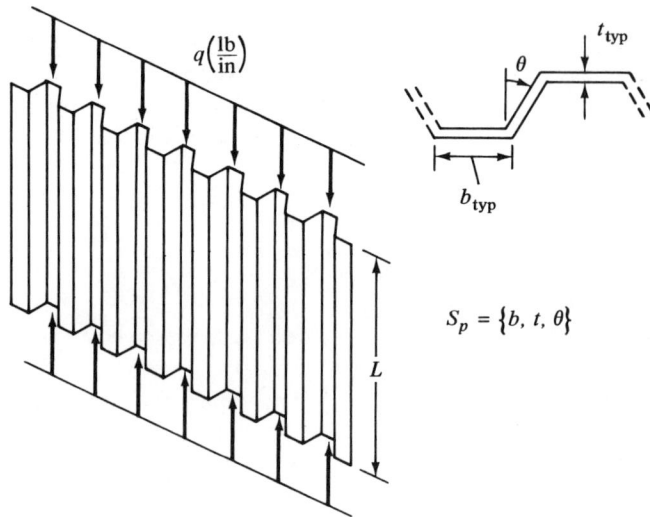

Fig. 7-1. Three-Variable Flat Corrugation Wide Column

The cross section is under uniform axial compression and hence the applied stress merit is applicable. The failure constraints are Euler buckling, local plate buckling of the flats, and excessive stress. In terms of the system environment and variables the local buckling constraint can be expressed as

$$\sigma_A = \psi_L 3.62 \eta_T^{1/2} E \left(\frac{t}{b}\right)^2 \tag{7-1}$$

$$\psi_L \leq 1$$

assuming pinned longitudinal restraint at the corrugation folds.

The Euler constraint is written first in terms of radius of gyration as

$$\sigma_A = \psi_E \frac{\pi^2 \eta_T E}{\left(\frac{cL}{r}\right)^2} \tag{7-2}$$

$$\psi_E \leq 1$$

(again in Eq. (7-2) cL represents an effective column length dependent on the condition of end support as given in Sec. 2-5).

To evaluate the radius of gyration we express the area of the representative width, w_R, as shown in Fig. 7-2, by

$$A_R = 4bt \tag{7-3}$$

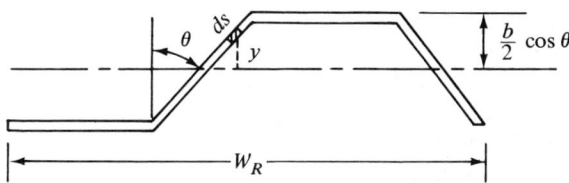

Fig. 7-2. Geometry of Representative Width

The moment of inertia of the two flats parallel to the buckling axis for the representative length can be expressed using a thin-walled approximation as

$$I_1 = 2bt \left(\frac{b}{2} \cos \theta\right)^2 \tag{7-4}$$

For the askew webs the moment of inertia can be obtained by integrating with respect to a differential length of flat over a half web and multiplying by four half webs as shown

$$I_2 = 4 \int_0^{b/2 \cos \theta} y^2 (t \, ds) \tag{7-5}$$

From the geometry shown in Fig. 7-2, ds can be written as

$$ds = \frac{dy}{\cos \theta} \tag{7-6}$$

Inserting Eq. (7-6) into Eq. (7-5) and integrating results in

$$I_2 = \frac{tb^3}{6} \cos^2 \theta \qquad (7\text{-}7)$$

Now combining Eqs. (7-4) and (7-7) yields

$$I_R = I_1 + I_2 = \frac{2}{3} tb^3 \cos^2 \theta \qquad (7\text{-}8)$$

The radius of gyration for the total width must be that for the representative width since both A_R and I_R increase by the same factor for any multiple of w_R. Therefore, employing Eqs. (7-3) and (7-8),

$$r^2 = \frac{I_R}{A_R} = \frac{b^2 \cos \theta}{6} \qquad (7\text{-}9)$$

Introducing Eq. (7-9) into Eq. (7-2) results in the Euler constraint in terms of the specific proportion variables as follows

$$\sigma_A = \psi_E \frac{\pi^2 \eta_T E b^2 \cos \theta}{6 c^2 L^2} \qquad (7\text{-}10)$$

The general approach for axially loaded elements requires algebraic combination of the buckling constraints, Eqs. (7-1) and (7-10), and an applied stress merit, σ_A/ρ, with the yield constraint providing an upper bound on applied stress. Since there are two buckling constraints and three proportion variables, we must express merit in terms of one "excess" proportion variable in addition to the slack variables. In general one or more ratios of proportion variables are selected as such excess variables since they often become secondary system variables. In the present case the angle θ itself is the appropriate variable to remain in the merit evaluation. There is no quantitative a priori basis for such a selection, however, exhaustive applications have indicated that corrugation angles and ratios of web and flange stiffener dimensions to the corresponding dimensions of the overall panel skin are normally secondary system variables.[13] We have already seen how this applies to the H-section as a candidate slender column or beam.

Proceeding on this basis we first express the applied stress merit in terms of proportion variables as follows

$$\frac{\sigma_A}{\rho} = \frac{P}{\rho A} = \frac{qw}{\rho A_R \left(\dfrac{w}{w_R}\right)} = \frac{qw_R}{\rho A_R} \qquad (7\text{-}11)$$

The representative width, w_R can be expressed from Fig. (7-2) as

$$w_R = 2b + 2b \sin \theta$$

Therefore, with A_R from Eq. (7-3)

$$\frac{\sigma_A}{\rho} = \frac{q(1 + \sin \theta)}{\rho 2t} \qquad (7\text{-}12)$$

Solving Eqs. (7-1) and (7-10) simultaneously for t (in terms of σ_A), and combining with Eq. (7-12) gives, after some algebraic manipulation, an expression for applied stress merit in terms of slack variables and θ as follows

$$\frac{\sigma_A}{\rho} = \psi_L^{1/4}\psi_E^{1/4}\left[\frac{\pi^{1/2}(3.62)^{1/4}}{c^{1/2}(24)^{1/4}}\right][(1 + \sin\theta)\cos\theta]^{1/2}\eta_T^{3/8}\frac{E^{1/2}}{\rho}\left(\frac{q}{L}\right)^{1/2} \quad (7\text{-}13)$$

subject to

$$\psi_L \leq 1,\ \psi_E \leq 1,\ \sigma_A \leq \sigma_y$$

We conclude therefore that $\psi_{L_{\mathrm{opt}}} = \psi_{E_{\mathrm{opt}}} = 1$ and hence the optimum cross section is an SMD when $\sigma_A \leq \sigma_y$. As in the case of a slender column, if the prediction of Eq. (7-13) exceeds σ_y, any selection of the product $\psi_L\psi_E$ which reduces Eq. (7-13) to σ_y will be optimum.

We see also that independent of material metric, $E^{1/2}/\rho$, and load index, q/L, the applied stress merit will be maximum when

$$f_e(\theta) = (1 + \sin\theta)\cos\theta \longrightarrow \max \quad (7\text{-}14)$$

Forming $df_e(\theta)/d\theta = 0$ results in

$$\theta_{\mathrm{opt}} = 30° \quad (7\text{-}15)$$

as a secondary system variable.

Inserting the optimum values of ψ_L, ψ_E and θ into Eq. (7-13) results in

$$\left(\frac{\sigma_A}{\rho}\right)_{\mathrm{opt}} = \frac{1.26}{c^{1/2}}\eta^{3/8}\frac{E^{1/2}}{\rho}\left(\frac{q}{L}\right)^{1/2} \quad (7\text{-}16)$$

The form of the material metric $E^{1/2}/\rho$ results in a ranking of materials such that magnesium is superior at the very low values of q/L, with aluminum and titanium becoming superior at the moderate and high load index ranges respectively due to the yield constraints. This comparison is shown in Fig. 7-3.

Geometric Constraint on Thickness for Flat Corrugation Wide Column Consider a constraint on t, $t \geq t_m$. Suppose that the load environment is such that $t_{\mathrm{opt}} < t_m$. We seek to evaluate ψ_L, ψ_E, θ, and b such that $t = t_m$ while minimizing the associated weight penalty. To evaluate t as a function of slack variables and the excess proportion variable θ, Eqs. (7-12) and (7-13) can be combined, yielding

$$t = \frac{(24)^{1/4}(qcL)^{1/2}}{2(3.62)^{1/4}\pi^{1/2}E^{1/2}}\frac{(1+\sin\theta)^{1/2}}{(\psi_E\psi_L)^{1/4}\cos^{1/2}\theta} \quad (7\text{-}17)$$

Now consider simultaneously the dependence of merit on ψ_L, ψ_E, and θ, viz.

$$\frac{\sigma_A}{\rho} \propto (\psi_E\psi_L)^{1/4}[(1+\sin\theta)\cos\theta]^{1/2} \quad (7\text{-}18)$$

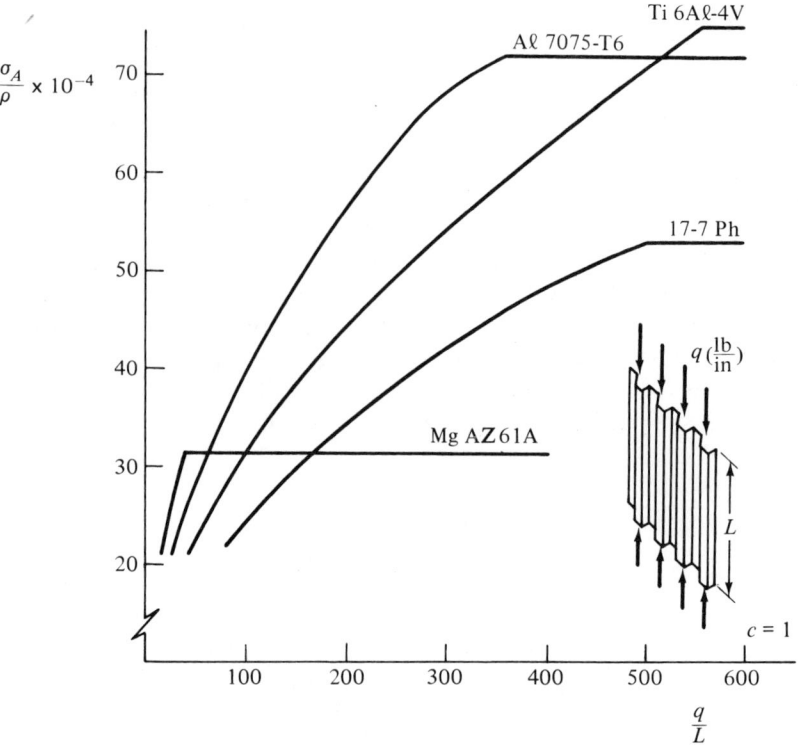

Fig. 7-3. Optimum Merit Function For Flat Corrugation Wide Column

Note that to introduce "slack" into either ψ_L or ψ_E, to increase t, results in a proportionate drop in merit. Hence to double t would cause the weight to double. The conclusion is that a conservative margin for either local or Euler buckling does not yield a desirable redistribution of material. To ascertain whether fluctuations in θ can achieve the constraint without having to accept a proportionate increase in weight, we can evaluate expressions for t_m/t_{opt} and $\sigma_{A\text{opt}}/\sigma_{A\text{co}}$ in terms of θ only from Eqs. (7-17) and (7-18) which results in

$$\frac{t_m}{t_{\text{opt}}} = \frac{\left(\dfrac{1+\sin\theta}{\cos\theta}\right)^{1/2}}{\left(\dfrac{1+\sin 30°}{\cos 30°}\right)^{1/2}} \tag{7-19}$$

$$\frac{\sigma_{A\text{opt}}}{\sigma_{A\text{co}}} = \frac{[(1+\sin 30°)\cos 30°]^{1/2}}{[(1+\sin\theta)\cos\theta]^{1/2}} \tag{7-20}$$

Equations (7-19) and (7-20) are shown plotted in Fig. 7-4.

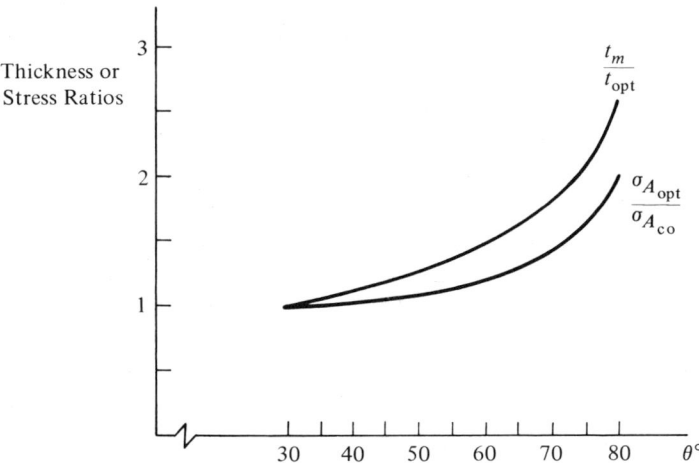

Fig. 7-4. Constrained Thickness and Stress Ratios as a Function of θ

Note in the figure that a geometric constraint which doubles t ($t_m/t_{opt} = 2$) results in a 57 percent increase in weight ($\sigma_{A_{opt}}/\sigma_{A_{co}} = 1.57$ for a constrained optimum value of θ of 74 degrees). The conclusion is that θ, although a secondary system variable in an open variable solution, becomes a primary variable dependent on t_m/t_{opt} in a constrained thickness solution.

Section 2
FRAME-STIFFENED CYLINDER IN BENDING

It was noted in Chap. 4 that a minimum-weight simple monocoque tube will carry a given bending moment at an optimum diameter which is normally too small to suffice for its use as a fuselage structure. Under a geometric constraint solution it was found that to increase the diameter would necessitate a conservative margin with respect to excessive stress. Although a redistribution of material allowed for a corresponding drop in thickness, the weight penalty was still excessive for large diameters.

In this section we will consider a more efficient solution to the large specified diameter shell under bending in the form of a circular tube stiffened both longitudinally and circumferentially so as to increase the stress for which the shell buckles (as shown in Fig. 7-5).

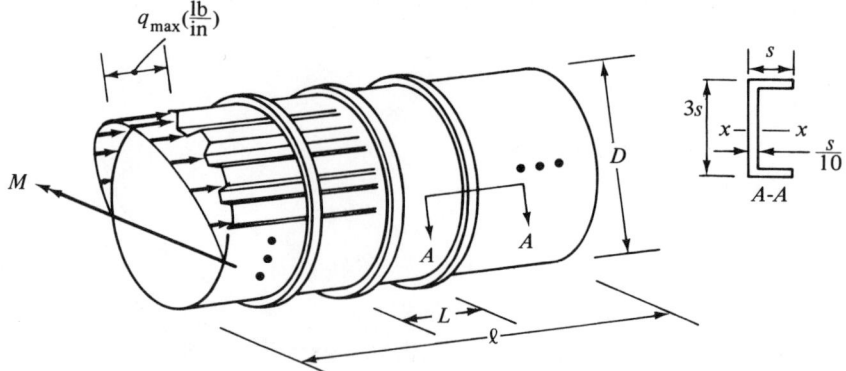

Fig. 7-5. Frame Supported Stiffened Skin Cylinder in Bending

We will leave the type of longitudinal stiffening general, employing an unspecified number of proportion variables. The corrugation shown in Fig. 7-5 is to be considered simply as representative, illustrating the orientation of the stiffening to the overall geometry. The function of the equally spaced frames is to provide support for the panel skin such that buckling occurs between frames. Since under the applied moment the frames are essentially unstressed,* the only failure constraint for these elements is that they be sufficiently "stiff" so as to enforce nodes for the skin panel. The condition for sufficient stiffness has been developed by Shanley[6] and can be expressed as a minimum requirement for the flexural stiffness (EI_f) of the frame cross section in terms of the frame spacing, L, applied moment, and tube diameter as follows

$$EI_f \geq C_f \frac{MD^2}{L} \tag{7-21}$$

where C_f is an empirically determined constant. Test data quoted in Ref. 6 indicates that a conservative value would be $C_f = 6.25 \times 10^{-5}$.

For a thin-walled frame cross section defined in terms of one proportion variable it is possible to relate the frame cross-sectional area, A_f, and moment of inertia, I_f, by

$$I_f = k_f A_f^2 \tag{7-22}$$

where k_f is a constant for a given frame defined in terms of one variable. To illustrate this, consider the typical cross section shown in Fig. 7-5. Here we find that

---

*With the exception of flexurally induced crushing loads which result from longitudinal curvature in the bent tube. These loads are minor and are normally ignored in the initial synthesis of configuration.

$$I_f = 2s\left(\frac{s}{10}\right)\left(\frac{3s}{2}\right)^2 + \frac{s}{10}\frac{(3s)^3}{12} = \frac{81}{120}s^4 \qquad (7\text{-}23)$$

and

$$A_f = 5s\left(\frac{s}{10}\right) = \frac{s^2}{2} \qquad (7\text{-}24)$$

and therefore

$$k_f = \frac{I_f}{A_f^2} = 2.70 \qquad (7\text{-}25)$$

Accordingly, combining Eqs. (7-21) and (7-22), the frame stiffness requirement can be expressed in terms of a minimum required cross-sectional area as follows

$$A_f \geq \left(\frac{C_f M}{k_f EL}\right)^{1/2} D \qquad (7\text{-}26)$$

Satisfaction of the failure constraint of Eq. (7-26) implies that the skin will buckle between frames. If we take provisionally that the optimum frame spacing is such that $L_{\text{opt}} < D$ (say 20 percent or less), it is reasonable to neglect the shell curvature and assume that the extreme fiber compression surface will buckle as a flat-panel wide column between frames. Under this assumption the configuration can be treated as a system combination of wide-column and frame elements with the frame spacing, L, recognized as an orientation variable.

For an overall length of cylinder, ℓ, the total system weight can be written as

$$W_s = N\rho A_f \pi D + \rho A_s \ell \qquad (7\text{-}27)$$

where $A_f \pi D$ is the volume of a single frame and N is the total number of frames. Neglecting end effects N can be expressed as

$$N = \ell/L \qquad (7\text{-}28)$$

Since the frames are non-load-carrying, there is no justification for providing more frame area than that required for stiffness. Hence the equality of Eq. (7-26) defines the expression for A_f. The appropriate expression for A_s in Eq. (7-27) must be the total skin cross section based on the maximum compressive stress at the extreme fibers. For a uniform cross section around the circumference we therefore must express A_s by assuming that the maximum applied load distribution, $q_{\max}$, acts uniformly around the entire circumference, yielding

$$A_s = \frac{P}{\sigma_{A_{\text{opt}}}} = \frac{q_{\max} \pi D}{\sigma_{A_{\text{opt}}}} \qquad (7\text{-}29)$$

If the skin cross section was specified as a flat corrugation, Eq. (7-16) would apply for $\sigma_{A_{opt}}$. To keep the wide column form unspecified we can express $\sigma_{A_{opt}}$ for the general wide column as[13]

$$\sigma_{A_{opt}} = KE^{1/2}\left(\frac{q_{max}}{L}\right)^{1/2} \tag{7-30}$$

where K depends on the type of skin cross section. Combining Eqs. (7-29) and (7-30) yields

$$A_s = \frac{\pi D q_{max}^{1/2} L^{1/2}}{KE^{1/2}} \tag{7-31}$$

To obtain q_{max} as a function of bending moment and geometry we first evaluate an expression for maximum applied stress employing the flexure equation, Eq. (4-1)

$$\sigma_{A_{max}} = \frac{M(D/2)}{I_s} \tag{7-32}$$

Without specifying the skin cross section, I_s can be expressed in terms of an "effective thickness" t_{eff}, defined simply as skin cross-sectional area per unit of circumferential length, as follows

$$I_s = \frac{\pi D^3 t_{eff}}{8} \tag{7-33}$$

Therefore from Eqs. (7-32) and (7-33)

$$\sigma_{A_{max}} = \frac{4M}{\pi D^2 t_{eff}} \tag{7-34}$$

By definition $q_{max} = \sigma_A t_{eff}$. Hence

$$q_{max} = \frac{4M}{\pi D^2} \tag{7-35}$$

Combining Eqs. (7-31) and (7-35) yields for A_s

$$A_s = \frac{2\pi^{1/2} M^{1/2} L^{1/2}}{KE^{1/2}} \tag{7-36}$$

We are now in a position to express total weight in terms of the orientation variable, L. Inserting Eqs. (7-26), (7-28), and (7-36) into Eq. (7-27) yields

$$W_s = \rho \ell \left(\frac{\pi C_f^{1/2} M^{1/2} D^2}{k_f^{1/2} E^{1/2} L^{3/2}} + \frac{2\pi^{1/2} M^{1/2} L^{1/2}}{KE^{1/2}}\right) \tag{7-37}$$

Since $M^{1/2}/E^{1/2}$ can be factored, we conclude that L_{opt} depends only on D and the constants which define frame and skin forms, k_f and K. Also by factoring $D^{1/2}$ a convenient orientation index (L/D) is introduced and results in

$$W_s = \rho \ell \left(\frac{MD}{E}\right)^{1/2} \left[\frac{C_f^{1/2}\pi}{k_f^{1/2}\left(\frac{L}{D}\right)^{3/2}} + \frac{2\pi^{1/2}}{K}\left(\frac{L}{D}\right)^{1/2}\right] \qquad (7\text{-}38)$$

Equation (7-38) illustrates an obvious trade-off process between skin and frames in terms of L. For small L/D the total frame weight dominates, whereas for large L/D the skin weight dominates. The values of L/D for which total weight attains a minimum can be obtained by evaluating $dW_s/d(L/D) = 0$ and yields (with $C_f = 6.25 \times 10^{-5}$)

$$\left(\frac{L}{D}\right)_{opt} = 0.145 \left(\frac{K^{1/2}}{k_f^{1/4}}\right) \qquad (7\text{-}39)$$

For example, taking a corrugation skin ($K = 1.26$) and the frame cross section of Fig. 7-5 ($k_f = 2.7$) results in $(L/D)_{opt} = 0.125$, which justifies our provisional assumption of $L_{opt} < D$. Figure 7-6 illustrates the weight sensitivity with respect to L/D for this case.

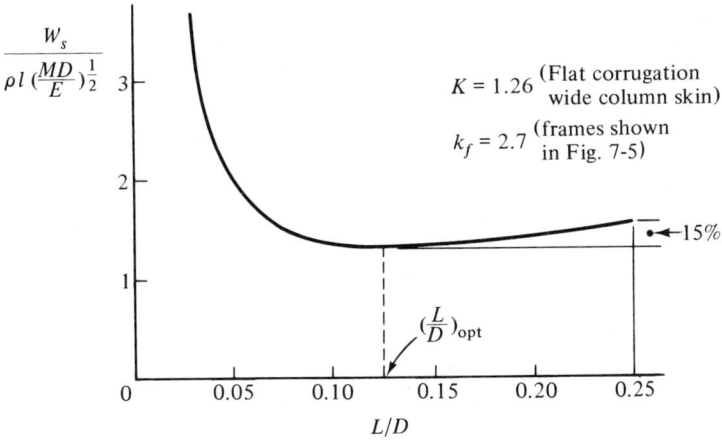

Fig. 7-6. Weight Sensitivity in Terms of L/D

Note the lack of sensitivity for $L/D > (L/D)_{opt}$. For example the frame spacing can be doubled with an increase in total weight of only 15 percent. If one accounts for fastening weight, such an increase would likely be justified.

Section 3

MINIMUM COST REINFORCED CONCRETE BEAM

One of the most widely used structural forms in civil engineering applications is the reinforced concrete beam. The relative cheapness of concrete makes it an ideal material for carrying compression stress where weight is not a factor. Unfortunately its tensile stress-carrying capability is nil and is assumed zero in practice. To use concrete as a beam cross section, therefore, requires steel reinforcing to carry the tensile load requirement. It is an empirical fact that the concrete can successfully transmit this tensile load to "knurled" steel reinforcing rods by shear traction and direct bearing against the knurls. In this section we will develop the equations for optimizing the proportions and the percentage of steel reinforcing based on the criterion of minimum cost.

Basically the heterogeneous concrete-steel composite can be considered a similar element system of axially stressed elements with differing material properties and unit volume costs. With this systems approach the percentage reinforcement can be treated as an orientation variable with specific proportions being dependent evaluations. A SMD (in this application referred to as a *balanced* design) is certainly not an intuitive solution to the minimum cost reinforced beam. Certainly the steel reinforcing, being the most expensive ingredient, will likely be fully stressed at the applied moment. To imply that the concrete will be fully stressed, however, does not intuitively lead to a minimum cost cross section. To provide extraneous concrete area and therefore an understressed state in the concrete could be justified from the standpoint of cost due to consequent reductions in required steel area resulting in a net cost savings. In what follows we will adopt the attitude that the stresses carried by both the steel and concrete are to be optimized, thus allowing for a possible understressed state in the concrete (in this application referred to as an *underreinforced* design).

Assuming linear behavior in the concrete, the maximum values of applied stress in the concrete and steel respectively can be expressed as[14]

$$\sigma_c = \frac{2M}{bd^2k\left(1 - \frac{k}{3}\right)} \tag{7-40}$$

$$\sigma_s = \frac{M}{A_s d\left(1 - \frac{k}{3}\right)} \tag{7-41}$$

where b, d, and A_s represent three proportion variables which define the rectangular dimensions and total steel cross section as shown in Fig. 7-7.

The parameter k in Eqs. (7-40) and (7-41) defines the neutral axis location as a fraction of d, and can be written in terms of the proportion and material variables as follows

$$k = \sqrt{N^2n^2 - 2Nn} - Nn \tag{7-42}$$

where

$$n = \frac{E_s}{E_c} \tag{7-43}$$

and
$$N = \frac{A_s}{bd} \tag{7-44}$$

Eq. (7-43) defines a moduli ratio constant for the steel-concrete composite whose value is on the order of 10. Equation (7-44) introduces the variable ratio of steel area to concrete area which is subject to optimization.

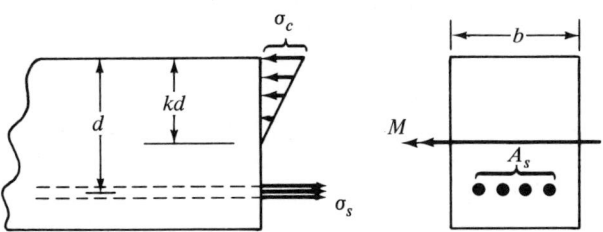

Fig. 7-7. Variable Description of Reinforced Concrete Beam

The cost per unit length of beam can be written in terms of the respective costs per unit volume for steel and concrete and cross-sectional areas as follows

$$C = c_s A_s + c_c bd \tag{7-45}$$

The failure constraints are excessive stress in the steel and concrete. Assuming *provisionally* that the concrete will be understressed at minimum cost, we express failure suppression simply as

$$\sigma_s = \psi_s \sigma_{sf} \tag{7-46}$$
$$\psi_s \leq 1$$

where σ_{sf} is the maximum permissible stress in the steel (a typical *working stress* value is 20,000 psi). Combining Eqs. (7-41) and (7-46) yields

$$\frac{M}{A_s d \left(1 - \frac{k}{3}\right)} = \psi_s \sigma_{sf} \tag{7-47}$$

Substituting $A_s = Nbd$ (Eq. (7-44)) into Eq. (7-47) and solving for d yields

$$d = \frac{M^{1/2}}{\psi_s^{1/2} \sigma_{sf}^{1/2} b^{1/2} \left(1 - \frac{k}{3}\right)^{1/2} N^{1/2}} \tag{7-48}$$

Now, since $A_s = Nbd$, Eq. (7-48) can be employed to express A_s as

$$A_s = \frac{M^{1/2} b^{1/2} N^{1/2}}{\psi_s^{1/2} \sigma_{sf}^{1/2} \left(1 - \frac{k}{3}\right)^{1/2}} \tag{7-49}$$

Inserting the above expressions for d and A_s into Eq. (7-45) yields for total cost

$$C = \frac{M^{1/2}b^{1/2}}{\psi_s^{1/2}\sigma_{sf}^{1/2}}\underbrace{\left[c_s\frac{N^{1/2}}{\left(1-\frac{k}{3}\right)^{1/2}} + c_c\frac{1}{N^{1/2}\left(1-\frac{k}{3}\right)^{1/2}}\right]}_{f_e(N)} \qquad (7\text{-}50)$$

subject to $\psi_s \leq 1$ and the proviso that the concrete will not be overstressed.

We see, therefore, that $\psi_{s_{\text{opt}}} = 1$, implying the steel will be fully stressed at minimum cost. Also since f_e in Eq. (7-50) is a function of N only for given unit costs and the ratio of moduli (recall that $k = f(N, n)$), it can be optimized independent of load environment. Note both d and A_s follow from Eqs. (7-48) and (7-49) for given N and load environment but that the beam width, b, cannot be optimized. Obviously b must be selected at some minimum value compatible with a maximum beam depth limitation since d depends inversely on b.

Since k in Eq. (7-50) is a complicated function of N, a numerical approach must be employed to evaluate N_{opt}, for given values of n, c_s, and c_c. Also, in the search for N_{opt} the provisional assumption that the concrete is understressed must be lifted, in that we must be able to guarantee that the value of N which minimizes cost corresponds to acceptable stress in the concrete. To effect such a guarantee, consider the ratio σ_c/σ_s, which by Eqs. (7-40) and (7-41) yields

$$\frac{\sigma_c}{\sigma_s} = \frac{2N}{k} \qquad (7\text{-}51)$$

and is thus seen to be a function of N only. Since the steel is fully stressed, Eq. (7-51) can be rewritten as

$$\sigma_c = \frac{2N}{k}\sigma_{sf} \qquad (7\text{-}52)$$

Hence the stress in the concrete can be monitored as N varies. Obviously the prediction of Eq. (7-52) at N_{opt} defines an optimum or ideal stress for the concrete. If the actual failure stress in the concrete, σ_{cf} (a typical working stress value would be 1300 psi) exceeded this value, the minimum cost cross section would be an underreinforced one. Figure 7-8 shows the cost efficiency factor, f_e, and optimum stress for the concrete plotted as a function of N. The cost efficiency factor was based on the "in place" (material plus installation) unit costs of $c_s = 0.15$ \$/lb (0.0425 \$/in³) and $c_c = 20$ \$/yd³ (0.000430 \$/in³). The corresponding optimum stress for the concrete was obtained from Eq. (7-52) at a value of $\sigma_{sf} = 20{,}000$ psi. Note that at the optimum percentage reinforcement ($N_{\text{opt}} = 0.91$ percent), the optimum stress in the concrete is only 980 psi. This value is considerably lower than the typical working stress value of 1300 psi and demonstrates that at the given unit cost figures, the optimum configuration will be underreinforced.

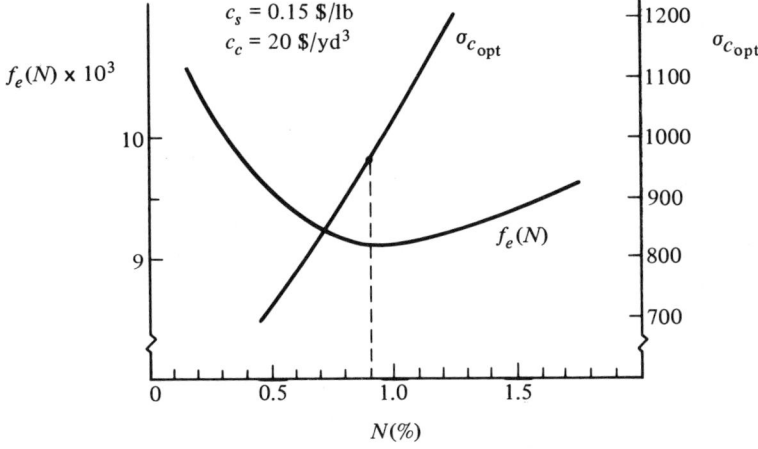

Fig. 7-8. Cost Efficiency Function and Optimum Concrete Stress in Terms of Percentage Reinforcement

Section 4
POST-BUCKLING COMPRESSION STRUCTURE

If we examine the various failure modes of structures, it is found that they can be classed either as defined or catastrophic. A *defined failure mode* is simply any event that does not constitute collapse of the structure. Limitations on deflection as well as the yield stress constraint are obvious examples of defined failure modes. The *catastrophic failure mode* is that which corresponds to physical collapse of the structure and is therefore inherently failure and need not be arbitrarily defined as such. The Euler buckling mode previously considered in slender and wide-column design is an obvious example of a catastrophic failure mode. But what of the various local buckling modes? Here we find that with one exception—the circular tube—all of the local buckling modes are actually defined failure modes. Local buckling of a flat or curved plate with longitudinal support created by web–flange intersections or corrugation folds does not represent catastrophic failure. In such a case a portion of the plate width adjacent to the longitudinal restraint remains unbuckled at the local buckling stress. This so-called *effective width* can support additional loads even when failure is defined as yield as long as local buckling occurs at a stress below yield stress. Designing to take advantage of this *post-buckling* phenomena does not necessarily represent an avenue of potentially significant weight savings. At the higher values of load index, local buckling occurs close to yield stress and therefore the post-buckling reserve is short-lived. Even at the lower values of load index the potential advantages of increasing stress beyond local buckling is offset by the fact that a buckled cross section suffers a loss of flexural rigidity and hence its effectiveness from the standpoint of Euler buckling or bending moment resistance is diminished. There is a lack of information on these offsetting effects for most cases. Gerard[15] presents an empirical interaction equation which allows for the assessment of the reduction in Euler allowable for a post-buckled wide column. Before considering a case as sophisticated as this, however, we will first abstract the flat plate element by ignoring the coupling effect with other failure modes and explore its weight advantage employing a post-buckling local failure mode.

Post-Buckling Flat Plate

Consider a simply supported plate as shown in Fig. 7-9.

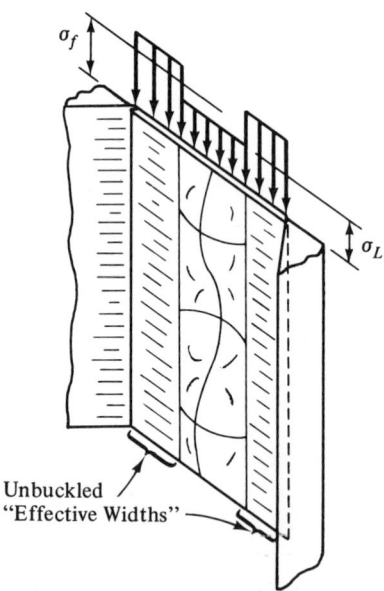

Fig. 7-9. Post-buckling of an Isolated Flat Plate

Beyond the local buckling stress the center portion of the plate is in a buckled state and cannot carry additional loads. The effective widths adjacent to the longitudinal edge supports continue to carry loads resulting in the idealized

Section 4 Post-Buckling Compression Structure

stepped stress distribution shown. The *average stress*, σ_f, for which ultimate local failure occurs in the post-buckled state can be expressed as[16]

$$\sigma_f = C_e(\sigma_y E)^{1/2}\left(\frac{t}{b}\right)^{3/4} \qquad (7\text{-}53)$$

where C_e is a constant dependent on edge support and material (for a simply supported aluminum plate, $C_e = 0.509$).

The applied average stress can be expressed in terms of the uniform load distribution, q, and plate thickness as follows

$$\sigma_A = \frac{P}{A} = \frac{qb}{tb} = \frac{q}{t} \qquad (7\text{-}54)$$

The post-buckling failure constraint can therefore be expressed from Eqs. (7-53) and (7-54) as

$$\frac{q}{t} = \psi_f C_e(\sigma_y E)^{1/2}\left(\frac{t}{b}\right)^{3/4} \qquad (7\text{-}55)$$

In an unstiffened plate, the minimum weight configuration will be obtained at minimum thickness. Therefore, under a post-buckling constraint, Eq. (7-55) can be solved for t, yielding

$$t = \frac{q^{4/7}b^{3/7}}{\psi_f^{4/7}C_e^{4/7}(\sigma_y E)^{2/7}} \qquad (7\text{-}56)$$

Obviously $\psi_{f_{opt}} = 1$, and dividing through by b results in a load index (q/b) formulation as follows

$$\left(\frac{t}{b}\right)_{PB} = \frac{1}{C_e^{4/7}(\sigma_y E)^{2/7}}\left(\frac{q}{b}\right)^{4/7} \qquad (7\text{-}57)$$

where the subscript PB denotes the minimum thickness index under a post-buckling constraint.

To compare this to the minimum weight configuration under a local buckling constraint, we first write the expression for local buckling

$$\sigma_L = 3.62\eta_T^{1/2} E \left(\frac{t}{b}\right)^2 \qquad (7\text{-}58)$$

Combining Eq. (7-58) with Eq. (7-54) yields

$$\frac{q}{t} = \psi_L 3.62 \eta_T^{1/2} E \left(\frac{t}{b}\right)^2 \qquad (7\text{-}59)$$

and solving for t

$$t = \frac{q^{1/3}b^{2/3}}{\psi_L^{1/3}(3.62)^{1/3}\eta_T^{1/6}E^{1/3}} \qquad (7\text{-}60)$$

Recognizing that $\psi_{L_{opt}} = 1$ and dividing through by b results in

$$\left(\frac{t}{b}\right)_{LB} = \frac{1}{(3.62)^{1/3}\eta_T^{1/6}E^{1/3}}\left(\frac{q}{b}\right)^{1/3} \qquad (7\text{-}61)$$

where the subscript LB denotes the minimum thickness index under a local buckling constraint.

Whether the plate local failure mode is defined as local buckling or post-buckling, we must establish the yield constraint as an overriding constraint. To do this we insist that

$$\sigma_A \leq \sigma_y \qquad (7\text{-}62)$$

which by Eq. (7-54) yields

$$\frac{q}{t} \leq \sigma_y \qquad (7\text{-}63)$$

or in index form

$$\left(\frac{t}{b}\right) \geq \frac{1}{\sigma_y}\left(\frac{q}{b}\right) \qquad (7\text{-}64)$$

Both Eqs. (7-57) and (7-61) are subject to the overriding constraint of Eq. (7-64) under a defined yield failure. These equations are shown plotted in Fig. 7-10 for Al 7075-T6 and simple edge supports.

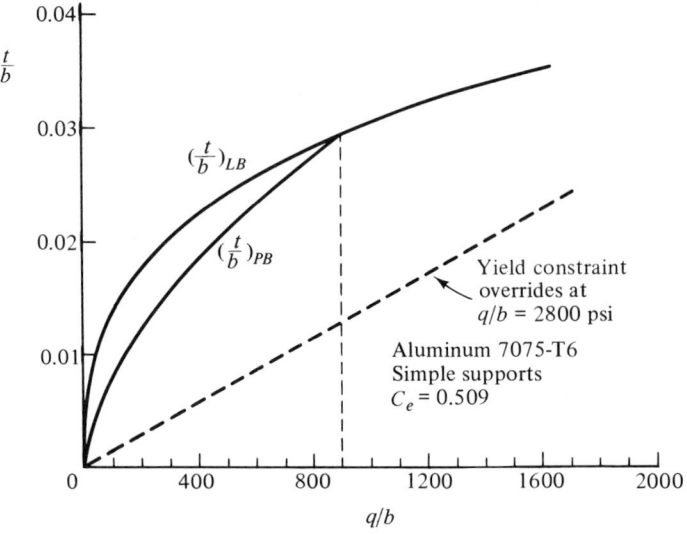

Fig. 7-10. Comparison of Local and Post-buckling Flat Plates for the General Environment

Note that below $q/b = 900$ psi the post-buckling design can represent significant weight savings. For example, at $q/b = 200$ psi the post-buckling design is approximately 40 percent lighter. Above $q/b = 900$ psi the local buckling design is lighter. This index is considerably below that for which yielding occurs ($q/b = 2800$ psi) and demonstrates that the post-buckling design represents a weight savings only at the relatively lower values of load index. It is interesting to note that if one were to *analyze* a cross section (designed on the basis of local buckling) using post-buckling failure criteria (Eq. (7-53)), the increased load-carrying capability would considerably misrepresent the potential weight savings that could be realized if post-buckling criteria were employed in the design of the cross section. Suppose that analyzing a local buckling design using Eq. (7-53) implies that X times the load could be carried before post-buckling failure. From the design viewpoint, however, we must ask how much the thickness can be reduced, not how much additional load can be carried. According to Eq. (7-59), to reduce the thickness causes the applied stress, q/t, to increase. A factor of increase of X, however, cannot be realized since the local buckling stress of the cross section—proportion to t^2—simultaneously becomes smaller. The net effect is illustrated in Eq. (7-60), where it can be seen that to design for applied stress exceeding the local buckling stress by a factor of X ($\psi_L = X$) results in a thickness reduction by a factor of only $X^{1/3}$.

Post-Buckling Flat Corrugation Wide Column

The weight advantage of the isolated post-buckling flat plate is significant enough at the lower values of load index to raise the question of whether a post-buckling local failure mode offers similar weight savings in more complex compression structures. First of all it should be noted that in high velocity aircraft a buckled surface may represent aerodynamic failure, and local buckling may be the defined failure mode regardless of the post-buckling weight advantage. Secondly, although it is reasonable to assume that a cross section on the verge of local buckling does not alter the cross section's flexural rigidity and hence does not diminish its resistance to "general" buckling (e.g. Euler buckling) and applied moment, it is unrealistic to suppose such an *uncoupled* behavior in the post-buckled state.

To illustrate the diminished returns that result from coupled general and local behavior, consider the flat corrugation wide column optimized in Sec. 1 under a defined local buckling failure mode. The expression developed there for applied stress merit, Eq. (7-13), although arrived at by manipulation of Euler and local buckling failure constraints, is nonetheless valid for the post-buckling state if we allow ψ_L to exceed unity, i.e., $\sigma_A > \sigma_L$. How much ψ_L can exceed unity can be answered by taking into account the post-buckling failure constraint, Eq. (7-53). The question of coupling with the Euler failure mode can be dealt with by evaluating the corresponding value of diminished ψ_E for the post-buckling value of ψ_L. Such an evaluation obviously requires a relationship which accounts for the coupling between the various failure modes in the post-buckled state. As previously noted in the introduction to this section, Gerard has developed and empirically verified a parabolic interaction equation for wide columns as shown in Fig. 7-11.

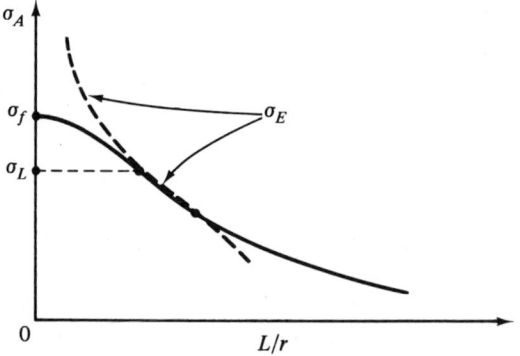

Fig. 7-11. Local, Euler and Post-buckling Interaction Curve

Note that if the applied stress is less than or equal to the local buckling stress of the cross section, there is no coupling since the applied stress can reach the Euler stress before failure. If the applied stress exceeds the local buckling stress, however, the stress at which failure occurs cannot reach the Euler stress and, in the limit, as σ_A continues to exceed σ_L, cannot exceed the post-buckling average failure stress, σ_f. The interaction equation which predicts a coupled applied failure stress, σ_A, for values of σ_L, σ_E, and σ_f can be written as[15]

$$\sigma_A = \sigma_f - [\sigma_f - \sigma_L]\frac{\sigma_L}{\sigma_E} \qquad (7\text{-}65)$$

Equation (7-65) can be employed to evaluate the diminished value of ψ_E for some $\psi_L > 1$ as a result of coupling. However, since this coupling also requires that $\sigma_A < \sigma_f$, we must introduce a slack variable, ψ_f. Introducing the slack variables ψ_L, ψ_E, and ψ_f into Eq. (7-65) and solving for ψ_E yields

$$\psi_E = \psi_L^2 \left[\frac{1 - \psi_f}{\psi_L - \psi_f}\right] \qquad (7\text{-}66)$$

Before ψ_E can be determined for a given value of ψ_L, a relation must be found for ψ_f in terms of ψ_L. This can be obtained by simultaneous consideration of the local buckling and post-buckling constraints for the corrugation flat-plate element as follows

$$\sigma_A = \psi_L 3.62 \eta_T^{1/2} E \left(\frac{t}{b}\right)^2 \qquad (7\text{-}67)$$

$$\sigma_A = \psi_f C_e (\sigma_y E)^{1/2} \left(\frac{t}{b}\right)^{3/4} \qquad (7\text{-}68)$$

Solving Eqs. (7-67) and (7-68) simultaneously to eliminate t/b yields

$$\psi_f = \frac{(3.62)^{3/8} \eta_T^{3/16} \sigma_A^{5/8}}{C_e \sigma_y^{1/2} E^{1/8}} \psi_L^{3/8} \tag{7-69}$$

Now combining Eqs. (7-66) and (7-69) yields

$$\psi_E = \psi_L^2 \left[\frac{1 - \phi \sigma_A^{5/8} \psi_L^{3/8}}{\psi_L - \phi \sigma_A^{5/8} \psi_L^{3/8}} \right] \tag{7-70}$$

where ϕ lumps the various constants as

$$\phi = \frac{(3.62)^{3/8} \eta_T^{3/16}}{C_e \sigma_y^{1/2} E^{1/8}} \tag{7-71}$$

Equation (7-70) predicts the coupled diminished value for ψ_E in terms of ψ_L at any applied stress level, σ_A. To evaluate the post-buckling advantage we can insert the above expression for ψ_E into the applied stress merit, Eq. (7-13), resulting in

$$\frac{\sigma_A}{\rho} = \psi_L^{3/4} \left[\frac{1 - \phi \sigma_A^{5/8} \psi_L^{3/8}}{\psi_L - \phi \sigma_A^{5/8} \psi_L^{3/8}} \right]^{1/4} \frac{\pi^{1/2} (3.62)^{1/4}}{(24)^{1/4}} [(1 + \sin \theta) \cos \theta]^{1/2}$$
$$\times \eta_T^{3/8} \frac{E^{1/2}}{\rho} \left(\frac{q}{L} \right)^{1/2} \tag{7-72}$$

Equation (7-72) cannot be solved explicitly for the applied stress merit, σ_A/ρ. However for a given σ_A the corresponding value of load index, q/L, can be evaluated explicitly. A numerical computer search for aluminum 7075-T6 yields the comparison shown in Fig. 7-12. By selecting values for σ_A, ψ_L can be optimized as that value for which the corresponding load index, q/L, is minimum. In this way the resulting plot, σ_A/ρ, in terms of q/L, will guarantee a maximum for applied stress merit for a given value of the load index. Note in Eq. (7-72) that for any σ_A and ψ_L, q/L will be minimum when $(1 + \sin \theta) \cdot \cos \theta$ is maximum. Hence $\theta_{opt} = 30°$ is seen to be a secondary system variable evaluation as was the case in the local buckling design considered in Sec. 7-1.

From the above figure it can be seen that the post-buckling design is optimum only at the very low values of load index ($q/L < 17$ psi corresponding to an applied stress range $\sigma_A < 16,400$ psi). Note that below this index the optimum value of ψ_L grows rapidly, reaching a value of 1.6 at a load index of 15. A comparison of post-buckling and local buckling designs at this index indicates only a 1.3 percent weight savings. As such the effects of general mode coupling can be seen to have almost totally offset the advantage gained by exceeding the local buckling stress of the cross section. Recall that in the isolated post-buckling flat plate the weight advantage for $\psi_L = 1.60$ would be $1.60^{1/3} = 1.17$, resulting in a 17 percent improvement over the local buckling design. In the present case the coupling phenomenon has further reduced this advantage to only 1.3 percent. Similar results are found for other wide column cross sections[19].

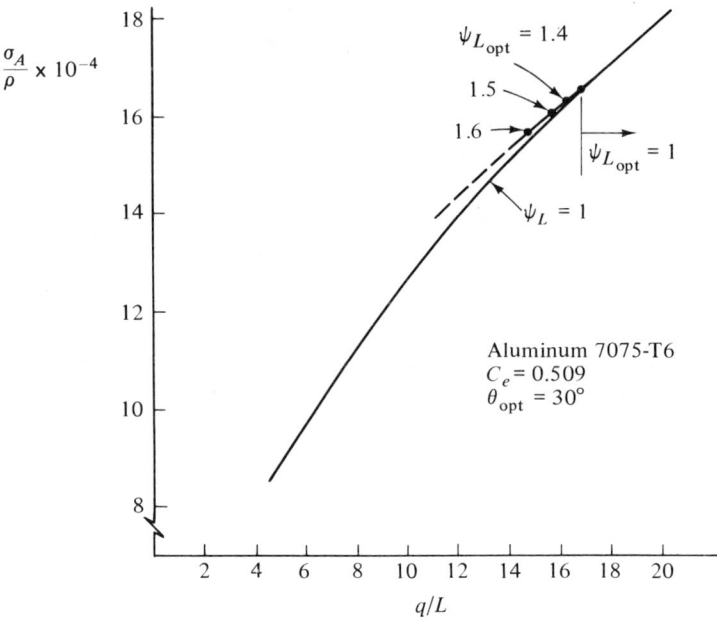

Fig. 7-12. Comparison of Local and Post-buckling Flat Corrugation Wide Columns for the General Environment

Unfortunately the parabolic interaction equation, Eq. (7-65), is not valid if the post-buckling failure stress, σ_f, exceeds twice the local buckling stress, σ_L. Since through coupling the applied stress cannot reach the post-buckling failure stress, it is found in the present case that $\psi_{L_{opt}} = (\sigma_A/\sigma_L)_{opt}$ must be less than 1.6 for a valid interaction. Accordingly the post-buckling advantage cannot be analytically evaluated below this point as indicated in Fig. 7-12 by the dashed line extrapolation below $q/L = 15$ psi.

The evaluation of an applied stress-dependent limitation for ψ_L can be effected by considering the geometrical limitation of the interaction curve as shown in Fig. 7-13. Note that the parabolic interaction curve must have two intersections (or a point of tangency) with the Euler allowable curve. The interaction equation forces one of these points of intersection to correspond to the local buckling stress. As σ_f exceeds σ_L a point is reached where the second intersection point σ_I, exceeds σ_L. This obviously invalidates the interaction expression since then the prediction will be that the applied stress could actually exceed the Euler allowable.

To establish the two intersection points, we can evaluate Eq. (7-65) for the points where $\sigma_A = \sigma_E = \sigma_I$, which yields

Post-Buckling Compression Structure

$$\sigma_I = \sigma_f[\sigma_f - \sigma_L]\frac{\sigma_L}{\sigma_I} \quad (7\text{-}73)$$

Rearranging gives the quadratic form

$$\sigma_I^2 - \sigma_f\sigma_I + (\sigma_f\sigma_L - \sigma_L^2) = 0 \quad (7\text{-}74)$$

which factors to

$$(\sigma_I - \sigma_L)(\sigma_I + \sigma_L - \sigma_f) = 0 \quad (7\text{-}75)$$

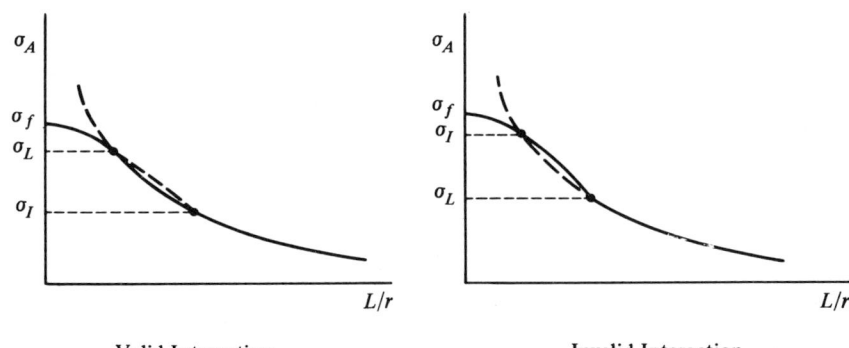

Valid Interaction   Invalid Interaction

Fig. 7-13. Limitation of the Interaction Equation

and results in the two roots $\sigma_{I_1} = \sigma_L$ and $\sigma_{I_2} = \sigma_f - \sigma_L$. The first root is the expected intersection at σ_L. The second root must be established as being less than σ_L for a valid interaction and yields the condition

$$\sigma_f - \sigma_L \leq \sigma_L \quad (7\text{-}76)$$

whence

$$\sigma_f \leq 2\sigma_L \quad (7\text{-}77)$$

which demonstrates that the post-buckling failure stress must be less than twice the local buckling stress for a valid interaction. Eq. (7-77) can be employed to solve for an upper bound value of ψ_L in terms of applied stress. Introducing slack variables into Eq. (7-77) yields

$$\psi_L \leq 2\psi_f \quad (7\text{-}78)$$

Employing Eq. (7-69), ψ_f can be expressed in terms of ψ_L and σ_A resulting in

$$\psi_L \leq \frac{(2)^{8/5}(3.62)^{3/5}\sigma_A}{C_e^{8/5}\sigma_y^{4/5}E^{1/5}} \quad (7\text{-}79)$$

Equation (7-79) was employed in the example of Fig. 7-12 to establish the region where the interaction equation was valid.

APPENDIX

Selected Material Properties

	Modulus of elasticity E (psi)	Density ρ (lb/in^3)	Yield stress σ_y (psi)		
Aluminum	10,000,000	0.100	2024-T3	2014-T6	7075-T6
			44,000	58,000	72,000
Steel	30,000,000	0.283	AISI 1025	17-7 PH Stainless	
			36,000	150,000	
Titanium	16,000,000	0.160	6Al-4V		
			120,000		
Magnesium	6,400,000	0.064	AZ61A		
			20,000		

Material Metrics

	Axial tensile members. Slender and wide columns at high load indices	Corrugation and other flat plate local element wide columns at low load indices	Square tube, H-section and other flat plate local element slender columns at low load indices	Circular tube column at low load indices	Rectangular tube and H-section beams	Circular tube beam
	σ_y/ρ	$E^{1/2}/\rho$	$E^{3/5}/\rho$	$E^{2/3}/\rho$	$E^{1/6}\sigma_y^{1/2}/\rho$	$E^{1/3}\sigma_y^{1/3}/\rho$
Aluminum 7075-T6	720,000	31,600	158,500	464,400	39,600	89,550
Steel 17-7 PH	530,000	19,350	108,200	341,400	24,260	58,280
Steel AISI 1025	127,000	19,350	108,200	341,400	11,900	36,220
Titanium 6Al-4V	750,000	25,000	131,300	397,000	34,550	77,600
Magnesium AZ61A	313,000	39,500	189,500	539,000	30,270	78,680

GLOSSARY

ACCEPTABLE CANDIDATES: Designs which satisfy the failure and geometric constraints.

ALGORITHM: A systematic numerical procedure where all decision outcomes are programmed.

ANALYSIS: The prediction of behavior employing a mathematical model of an idealized system.

ANALYTICAL OPTIMIZATION: A parametric algebraic solution to a synthesis evaluation.

APPLIED STRESS MERIT: Equivalent to the weight merit for centric axially loaded elements. Expressed as the applied stress divided by the materials weight density, a maximum implies minimum weight (σ_A/ρ).

AUTOMATED DESIGN: A numerically directed search method which employs a systematic algorithm for evaluating an optimum design.

CONSERVATIVE DESIGN: A design for which a margin exists against one or more failure modes.

CRITERIA: The basis for which alternative acceptable candidates can be ranked in terms of relative desirability. Generally expresses both quantitative factors such as cost or weight (or a trade-off) and qualitative factors such as aesthetics.

DESIGN SPACE: A mathematical space, the axes of which correspond to the design variables. Constraints divide such a space into acceptable and unacceptable design regions with the merit function establishing contours of constant merit.

DETERMINATE SYSTEM: A structural system for which the loads carried by each of the component elements can be evaluated from the equations of equilibrium.

EFFICIENCY FACTOR (or function): A function of system variables which can be optimized independent of the magnitude of the load environment.

ELEMENT: A component part of a structural system.

EXCESS VARIABLES: Proportion variables which remain in the merit function after algebraic elimination of design variables by employing the failure constraints. Generally excess variables are secondary system variables in open-variable design.

FAILURE CONSTRAINT: An inequality in terms of the system variables which expresses the suppression of a mode of failure.

FORM: A parameter definition of the shape of a structural element or system of elements.

GEOMETRIC CONSTRAINTS: Performance or fabrication considerations which express minimum and/or maximum values for specific design variables.

ITERATIVE DESIGN: A cyclic method in which repeated guessing of design variables becomes more educated based on the analysis of prior trials.

LOAD INDEX: A lumped representation of the system environment. Generally with the units of psi.

LOCAL ENVIRONMENT: The loads and transmission lengths associated with a given component element of a system of elements (E_{si}).

MATERIAL METRIC: A lumped representation of the relative efficiency of a material for some given load transmission function, expressed in terms of the material properties.

MATERIAL VARIABLES: Quantities which express the properties of a material (S_m).

MATHEMATICAL PROGRAMMING: A branch of mathematics concerned with the development of algorithms for solving inequality constrained optimization problems.

MERIT FUNCTION: A function which quantitatively expresses the principal part of the criteria (M).

OPEN-VARIABLE DESIGN: A design solution exclusive of any geometric constraints.

ORIENTATION VARIABLES: Variables which define the relative arrangement of a system of elements (S_o).

PROPORTIONS: Quantities which define size for a structural element or a system of elements. Incorporates both proportion variables and orientation variables.

PROPORTION VARIABLES: Variables which define size for a component element (S_p).

PRIMARY VARIABLES: Variables whose optimum values depend on a numerical expression of the load environment.

REDUNDANT SYSTEM: A statically indeterminate structural system.

REDUNDANT VARIABLE: A variable which expresses the load state of a redundant element or support. Typically defined as a fraction of an applied load (k).

SECONDARY SYSTEM VARIABLES: Variables whose optimum values can be evaluated independent of the magnitude of the system environment.

SENSITIVITY ANALYSIS: Establishes the variation in merit for variations in a given system variable away from the optimum.

SIMULTANEOUS MODE DESIGN: A design technique which equates failure modes as a method for evaluating an optimum design (SMD).

SLACK VARIABLES: Variables introduced into inequality constraints in order to convert them to equalities (ψ).

SYNTHESIS: The rational determination of the variables which define a system based on given criteria and subject to inherent and given constraints.

SYSTEM ENVIRONMENT: The quantitative parameters of a structural load environment such as loads and transmission lengths (E_s).

SYSTEM VARIABLES: Incorporates the proportion, material, and orientation variables (S).

TRADE-OFF: A compromise between conflicting criteria.

TRUSS SYSTEM: A system of straight elements mutually pinned at the ends and loaded at pin locations only.

VALUE COMPARATOR: The basis for effecting a trade-off. In a cost-weight trade-off it is the dollars you are willing to spend to save a pound of weight (V).

REFERENCES

1. Michell, A. G. M. "The limits of economy of material in frame-structures." *Phil. Mag.* Series 6, Vol. 8, No. 47, (1904), 589–597.
2. Maxwell, C. *Scientific Papers II.* (Cambridge University Press, 1890) pp. 175–177.
3. Asimow, Morris. *Introduction to Design.* (Englewood Cliffs, New Jersey: Prentice-Hall, 1964.)
4. Schmit, Lucien A. "Automated design." *International Science and Technology* (1966).
5. Shanley F. R. "Principles of structural design for minimum weight." *J. Aero. Sci.* Vol. 16, No. 3 (1949).
6. Shanley, F. R. *Weight-Strength Analysis of Aircraft Structures.* (New York: McGraw-Hill, 1952.)
7. Wilde, D. J. and Beightler, C. S. *Foundations of Optimization.* (Englewood Cliffs, New Jersey: Prentice-Hall, 1967.)
8. Rubinstein, M. F., and Karagozian, J. "Building design using linear programming. *Proc. ASCE* 92, ST 6 (1966), 223–245.
9. Schmit, L. A., Kichner, T. P., and Morrow, W. M. "Structural synthesis capability for integrally stiffened waffle plates." *AIAA J* 1 (1963), 2820–2836.
10. Feigen, M. "Minimum weight of a tapered round thin-walled column," *J. Appl. Mech.* 19 (1952), 375-380.
11. Timoshenko, S. P., and Gere, J. M. *Theory of Elastic Stability.* (New York: McGraw-Hill, 1961.)
12. Emero, D. H., and Alvey, V. W. "Structures cost effectiveness." *Proceedings of the AIAA 4th Aerospace Sciences Meeting*, Los Angeles, Calif., June 1966.
13. Emero, D. H. and Spunt, L., "Wing box optimization under combined shear and bending." *AIAA J. of Air.*, Vol. 3, No. 2 (1966), 130–141.
14. Dunham, C. *The Theory and Practice of Reinforced Concrete.* (New York: McGraw-Hill, 1966.)

15. Gerard, G. "Handbook of structural stability, part v: Compressive strength of flat stiffened panels," *NACA* TN3785 (1957).
16. Needham, R. A. "The ultimate strength of aluminum-alloy formed structural shapes in compression," *J. Aero. Sci.* (1954).
17. Sved, G., "The minimum weight of certain redundant structures," *Aust. J. Appl. Sci.*, 1954.
18. Cox, H. L., *The Design of Structures of Least Weight*, Pergamon Press, London, 1965.
19. Spunt, L. "Weight optimization of the postbuckled integrally stiffened wide column." "*AIAA J. of Air.* Vol. 7, No. 4 (1970) 330–333.

INDEX

INDEX

A

Acceptable design, 12
Analysis, 1
Analytical optimization, 17
Applied stress merit, 36
Automated design:
 definition of, 14, 20
 random steps, 22–23
 steepest descent, 22

B

Beam:
 circular tube, 64
 geometric constraints, 72
 H-section, 67
 reinforced concrete, 143
 tapered, 73
Bending stress:
 slender beam, 64
Buckling coefficient:
 beam web, 68
 circular tube, 36
 plates, 41

C

Catastrophic failure, 147
Circular tube beam, 64
Column:
 circular tube, 35
 geometric constraints, 49
 H-section, 40
 slender, 32
 wide, 133
Compatibility in redundant
  structures, 78, 115
Concrete beams:
 minimum cost of, 143
 reinforced design of, 144
 underreinforced design of, 144
Conservative margin, 72
Constrained optimization:
 circular tube beam, 72
 circular tube column, 49
Constraints:
 failure, 12
 geometric, 12, 29
 recognition of, 13
Cost-weight trade-off design, 52
Cost-weight trade-off merit
  function, 55

Cox, 130
Criteria:
 definition of, 9
 establishment of, 8
 qualitative aspect of, 10

D

Defined failure mode, 147
Design:
 automated, 14
 criteria establishment, 8
 hyperspace, 20
 iterative, 6
 phases of, 7
 process of, 6
 rational approach to, 2
 simultaneous mode, 15
 space, 12, 20, 25
 structural system, 76
Determinate system, 77
 dissimilar element, 87
 similar element, 81
Dissimilar element redundant system, 109
Dollar value of a pound, 53

E

Effective width of plate, 147
Emero, 56
End supports for columns, 49
Environment:
 factors of, 7
 local, 79
 systems, 7
Euler buckling, 32

Excess proportion variables, 29
Excessive stress or yield stress, 36

F

Failure constraints, 12
Failure modes:
 catastrophic, 147
 defined, 147
 Euler buckling, 35
 excessive stress, 36
 local buckling, 35
Fixed orientation, 85
Flat configuration wide column, 137
Flexural stress, 64
Form:
 optimization of, 6
 specification of, 10
 structural, 9
Frame-stiffened cylinder in bending, 139

G

Geometric constraints, 12, 29
 circular tube beam, 72
 circular tube column, 49
 flat corrugation wide column, 137
 systems, 97
Gerard, 148

H

H-section beam, 67
H-section column, 40
 web and flange buckling, 42

INDEX

I

Index:
  load, 38
  material, 38
  weight, 45
Inelastic correction, 36
Iterative design, 6

L

Lagrangian multiplier techniques, 15
Load index, 38, 90
Local buckling, 34
Local environment, 79

M

Material metric, 38
Material variables, 11
Mathematical programming, 14
Maxwell, 5
Merit function:
  cost-weight trade-off, 55
  definition of, 9, 54
Michell, 5, 130
Multiple load conditions, 97, 118, 126

N

Nature of structural design, 4

O

Open variable, 12
Optimization:
  axial force transmission, 5
  phase of, 13
  slack variables, 16
Optimum stress, 38
Optimum structural design:
  definition of, 9
Orientation variables, 11, 76

P

Plate buckling mode, 41
Point solution, 14
Post buckling:
  of compression structure, 147
  of flat corrugation wide column, 151
  of flat plate, 148
Pressurized cylinder, thickness of, 3
Prestressed:
  redundant systems, 116, 118, 120, 126
  structures, 109
Primary variables, 11
Proportions of structures, 4
Proportion variables, 6

R

Radius of gyration, 35
Random steps, 22
Recognition of constraints, 12
Redundant:
  dissimilar element system, 110
  external restraint, 102
  prestressed system, 118

Redundant (*cont.*):
  similar element system, 105
Reinforced concrete beam, 143

S

Schmit, L. A., 127
Secondary variables, 11
Sensitivity analysis, 84
Shanley, F. R., 15
Simultaneous mode design, 15
Slack variables, 16
Slender:
  beam, 63
  column, 32
Statically indeterminate structures, 102
Steepest descent, 22
Structural:
  element, 11
  index, 15
Structure:
  configuration of, 5
  definition of, 5
  proportion of, 5
  shape of, 5
Subsystem, 6, 25
Sved, 124
Synthesis, 1
System:
  environment, 7
  variables, 11
System design:
  analytical approach, 80
Systems:
  determinate, 77
  redundant, 77
System types:
  composite, 78
  dissimilar elements, 78
  prestressed, 78

System types (*cont.*):
  similar elements, 78, 87

T

Tapered beam, 73
Tensile load transmission, 17
Trade-off, 9, 52
Truss:
  definition of, 90
  Michell optimization of, 5, 130
  multiple load conditions, 118
  prestressed, 120, 126

V

Value comparator, 52
Variables:
  excess proportion, 29
  material, 11
  open, 12
  orientation, 11
  primary, 11
  proportion, 11
  secondary, 11
  slack, 16
  system, 11

W

Weight-cost trade-off design, 52
Weight index, 45
Wide column:
  definition of, 133
  flat corrugation, 134
  geometric constraints, 137

TA
658
S68

DEC 5 1973